Visualizing Quantum Mechanics with Python

Quantum Mechanics can be an abstract and complex subject. Students often complain of confusion, struggle, and frustration as they try to master the topic. The goal of this book is to reduce the complexity and clarify the abstractions with concrete visual examples driven by simple python programs. It is assumed that the reader is concurrently taking a course in quantum mechanics, or self-studying quantum mechanics, but is looking for supplementary material to help with understanding and visualizing how quantum mechanics works.

The focus of this book is writing python programs to visualize the underlying behavior of the mathematical theory. The background needed to understand quantum mechanics is differential equations, linear algebra and modern physics. We need a strong foundation in differential equations and linear algebra because the behavior of quantum systems is governed by equations that are written in terms of these concepts. Modern physics includes concepts such as special relativity and quantum phenomena like the photoelectric effect and energy quantization that the theory of quantum mechanics seeks to explain. This book is also not an introduction to the python programming language, or to numpy, or even to VPython. However its programming examples start simply and grow more complex as the chapters progress, so deep expertise in any of these is not a pre-requisite.

Key features:

- Provides an accessible and practical guide to the abstractions in quantum mechanics with concrete visual examples driven by simple python programs.
- Contains few derivations, equations, and proofs.
- For complete beginners of python programming, appendix B serves as a very brief introduction to the main concepts needed to understand the code in this book.

Dr. Steve Spicklemire is Associate Professor of Physics at the University of Indianapolis, USA. He has been teaching physics at the University of Indianapolis for more than 30 years. From the implementation of "flipped" physics class to the modernization of scientific computing and laboratory instrumentation courses, he has brought the strengths of his background in physics, engineering and computer science into the classroom. Dr. Spicklemire also does IT and engineering consulting. He is an active participant in several national research initiatives relating to improving physics education. These range from improving materials to help students prepare for class, to supporting students success through standards based grading. He is an active developer of the VPython and WebVPython projects and a contributor to the Matter and Interactions textbook.

Visualizing Quantum Mechanics with Python

Steve Spicklemire

CRC Press
Taylor & Francis Group
Boca Raton London New York

CRC Press is an imprint of the
Taylor & Francis Group, an **informa** business

First edition published 2024
by CRC Press
2385 NW Executive Center Drive, Suite 320, Boca Raton FL 33431

and by CRC Press
4 Park Square, Milton Park, Abingdon, Oxon, OX14 4RN

CRC Press is an imprint of Taylor & Francis Group, LLC

Library of Congress Cataloging-in-Publication Data

Names: Spicklemire, Steve, author.
Title: Visualizing quantum mechanics with Python / Steve Spicklemire.
Description: First edition. | Boca Raton, FL : CRC Press, 2024. | Includes
bibliographical references and index. | Summary: "Quantum Mechanics can
be an abstract and complex subject. Students often complain of
confusion, struggle, and frustration as they try to master the topic.
The goal of this book is to reduce the complexity and clarify the
abstractions with concrete visual examples driven by simple python
programs. It is assumed that the reader is concurrently taking a course
in quantum mechanics, or self-studying quantum mechanics, but is looking
for supplementary material to help with understanding and visualizing
how quantum mechanics works. The focus of this book is writing python
programs to visualize the underlying behavior of the mathematical
theory. The background needed to understand quantum mechanics is
differential equations, linear algebra and modern physics. We need a
strong foundation in differential equations and linear algebra because
the behavior of quantum systems is governed by equations that are
written in terms of these concepts. Modern physics includes concepts
such as special relativity and quantum phenomena like the photoelectric
effect and energy quantization that the theory of quantum mechanics
seeks to explain. This book is also not an introduction to the python
programming language, or to numpy, or even to VPython. However its
programming examples start simply and grow more complex as the chapters
progress, so deep expertise in any of these is not a pre-requisite"--
Provided by publisher.
Identifiers: LCCN 2023054393 | ISBN 9780367768799 (hbk) | ISBN
9781032569246 (pbk) | ISBN 9781003437703 (ebk)
Subjects: LCSH: Quantum theory--Data processing. | Python (Computer program
language)
Classification: LCC QC174.17.D37 S65 2024 | DDC
530.120285/5133--dc23/eng/20240209
LC record available at https://lccn.loc.gov/2023054393

ISBN: 978-0-367-76879-9 (hbk)
ISBN: 978-1-032-56924-6 (pbk)
ISBN: 978-1-003-43770-3 (ebk)

DOI: 10.1201/9781003437703

Typeset in Latin Roman font
by KnowledgeWorks Global Ltd.

Publisher's note: This book has been prepared from camera-ready copy provided by the authors.

Contents

Acknowledgment

I would like to thank my colleagues Ruth Chabay, Bruce Sherwood, and Aaron Titus for their unwavering support and encouragement and for creating WebVPython and the Matter and Interactions textbook which were the greatest influences on this project. I'd like to thank my editors, Danny Kielty and Ashraf Reza for their (nearly) inexhaustible patience, and helpful feedback. I'd like to thank my loving wife, Tawn Spicklemire, for always supporting me in my various crazy endeavors.

## CHAPTER 1

Introduction

Q UANTUM MECHANICS can be an abstract and complex subject. Students often complain of confusion, struggle, and frustration as they try to master the topic. The goal of this book is to reduce the complexity and clarify the abstractions with concrete visual examples driven by simple python programs. It is assumed that the reader is concurrently taking a course in quantum mechanics, or self-studying quantum mechanics, but is looking for supplementary material to help with understanding and visualizing how quantum mechanics works. There are few derivations, equations, or proofs. Those are left to the more formal textbooks and lectures. The focus of this book is writing python programs to visualize the underlying behavior of the mathematical theory. The background needed to understand quantum mechanics is differential equations, linear algebra and modern physics. We need a strong foundation in differential equations and linear algebra because the behavior of quantum systems is governed by equations that are written in terms of these concepts. Modern physics includes concepts such as special relativity and quantum phenomena like the photoelectric effect and energy quantization that the theory of quantum mechanics seeks to explain. This book is also *not* an introduction to the python programming language, or to numpy, or even to VPython. Having said that, the programming examples will start simply and grow more complex as the story line progresses, so deep expertise in any of these is not a pre-requisite. If you're completely new to python programming, appendix B has been provided as a very brief introduction to the main concepts needed to understand the code in this book.

1.1 VISUALIZING COMPLEX NUMBERS: THE COMPLEX PLANE

When we say quantum mechanics is *complex*, it's meant quite literally. The underlying theory of quantum mechanics is based on complex numbers: numbers with both a real and imaginary part, like so:

$$z = a + ib \tag{1.1}$$

You can think of a complex number z as having two components, a a *real part* and b an *imaginary part*. These two parts behave a little bit like coordinates in a two-dimensional cartesian space called the *complex plane*. This can be visualized as a spatial coordinate system as illustrated in Fig. 1.1.

1.2 PHASORS

Another way to visualize a complex number is as shown in Fig. 1.2 as a phasor. A phasor has a magnitude R and a phase θ.

DOI: 10.1201/9781003437703-1

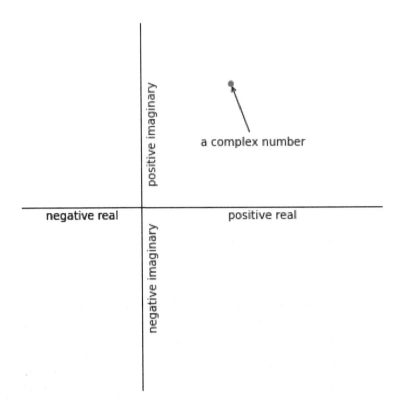

Figure 1.1   The complex plane.

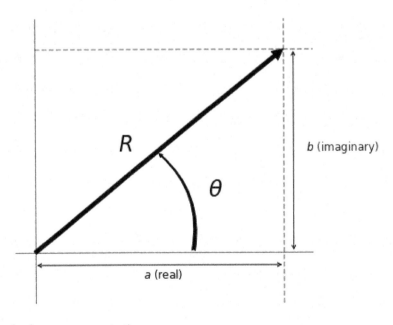

Figure 1.2   A phasor representation.

If you recall the Euler formula:

$$e^{i\theta} = \cos\theta + i\sin\theta \tag{1.2}$$

you can immediately see that

$$Re^{i\theta} = R\cos\theta + iR\sin\theta = a + ib \tag{1.3}$$

This way of expressing a complex number using the Euler formula is sometimes called *polar* form or *exponential* form. So, a and b can be thought of as the real and imaginary *components* of the *phasor z* where:

$$a = R\cos\theta \tag{1.4}$$
$$b = R\sin\theta \tag{1.5}$$

1.3 EXAMPLES

1.3.1 Example: Convert a Number from Rectangular to Exponential Form

Suppose you're given a complex number: $z = 3 + 4i$. What is its exponential form?

1.3.1.1 *Paper Solution*

Since we have a and b we can treat these almost like the x and y components of a *vector*:

$$R = \sqrt{a^2 + b^2} = \sqrt{9 + 16} = 5 \tag{1.6}$$

Similarly, we can find the angle using trigonometry:

$$\theta = \tan^{-1}\left(\frac{b}{a}\right) = \tan^{-1}\left(\frac{4}{3}\right) \tag{1.7}$$

So we get the final answer:

$$z = 5e^{i\theta} \text{ where } \theta = \tan^{-1}\left(\frac{4}{3}\right) \tag{1.8}$$

1.3.1.2 *Python Solution*

We can also do these calculations in python:

```
>>> z = 3 + 4j
>>> R = abs(z)
>>> print(R)
5.0
```

Note that you don't have to compute the square root of the sum of squares manually in python, you can just use the **abs** function to get the "absolute value" or magnitude of the complex number. Similarly, to get the angle we can use the **math.atan2** function to compute the angle. The **np.arctan2** function takes the y, and x components (respectively) of a vector and returns the angle the vector makes with the $+x$ axis. Of course it also works for phasors.

```
import numpy as np
theta = np.arctan2(z.imag, z.real)
print(np.degrees(theta))
53.13010235415598
```

1.3.2 Example: Convert a Number from Exponential to Cartesian Form

Suppose you're given a complex number: $z = 2e^{i\frac{\pi}{6}}$. What is its cartesian form?

1.3.2.1 Paper Solution

Since we have R and θ we can treat these almost like the magnitude and direction of a *vector*:

$$
\begin{aligned}
a &= R\cos\theta &= 2\cos(\pi/6) = 2\sqrt{3}/2 = \sqrt{3} & \quad (1.9)\\
b &= R\sin\theta &= 2\sin(\pi/6) = 2(1/2) = 1 & \quad (1.10)
\end{aligned}
$$

So we get the final answer:

$$
z = 2e^{i\frac{\pi}{6}} = \sqrt{3} + i \quad (1.11)
$$

1.3.2.2 Solution in Python

Once again, this can also be done in python:

```
R = 2
theta = np.pi/6
z = R*np.exp(1j*theta)
print(z)
(1.7320508075688774+0.9999999999999999j)
```

1.4 COMPLEX FUNCTIONS OF REAL NUMBERS

So, you recall *complex number* has a real part (a) and an imaginary part (b), or equivalently, a magnitude (R) and a phase (θ). Remember that a *function* is a kind of recipe that converts arguments (inputs) into a value (output). Quantum mechanics is built around special functions, called *wavefunctions* that have *real* arguments (e.g., position, x, or time, t), but whose value (output) is a *complex* number. One of the most frequently encountered complex functions in quantum mechanics is the complex exponential, like so:

$$
\phi(t) = Ae^{-i\omega t} \quad (1.12)
$$

In most cases, ω and t are *real* numbers and contribute only to the *phase angle* of the complex number. If we think of $\phi(t)$ as a phasor, it would be an arrow of fixed length A rotating clockwise in the complex plane with an angular velocity of ω radians per second.

We can also visualize a complex function by creating a 3D graphical representation in VPython. We'll let time be represented by a *real* direction in space, and for every time we'll draw a phasor that represents the complex value of the function at that particular time. See Fig. 1.3. Note how we can use a 3D spatial visualization to capture the idea of a complex function of a real variable. This is one of the *core* ideas of this book. In order to "see" a concrete representation of this complex function, we can use 3D space to visualize the

Figure 1.3   A 3D complex function.

phasors in 2D and then use the remaining dimension to picture how the function depends on time.

Another common case is a function created by multiplying a function of position by a function of time, like so:

$$\psi(x,t) = Ae^{ikx}e^{-i\omega t} \tag{1.13}$$

This function produces a phasor output that depends on *two* real variables, both t (time) and x (position). Unfortunately, it's not very easy to represent this kind of function as a static picture on a page, although we'll do our best in the next chapter. However, you *can* visualize this kind of complex function of space and time quite easily through animation! Another major theme of this book is to learn how to create 3D visual animations of complex wavefunctions to improve your conceptual understanding of quantum mechanics. The idea is to represent space as a one physical dimension in 3D (e.g., the x direction), the real and imaginary parts of the wavefunction value as two other dimensions in 3D (e.g., the y and z directions), and to get the time, one can animate the 3D figure in time using a computer program. You can see this example by scanning the QR code from Fig. 1.3. You'll see many examples of static images in this book accompanied by QR codes that can be used to see the actual animation as well as the code behind it.

1.5   MODERN PHYSICS CONCEPTS: PLANCK, EINSTEIN, DE BROGLIE, AND HEISENBERG

By the time you take up the subject of quantum mechanics it's likely that you've already encountered some of the basic ideas of Modern Physics. This section is a quick review of two important concepts we'll need right away to understand quantum wavefunctions.

1.6   THE PLANCK-EINSTEIN RELATIONSHIP

Planck used quantum concepts in the context of thermodynamics to develop a theory of blackbody radiation that fit experimental observation. He postulated that the vibrating atoms in the walls of the blackbody could only emit or absorb specific "quantized" amounts of energy proportional to the frequency of vibration. Through this mechanism he could explain why the short wavelength, high frequency waves, in the blackbody could not contain unlimited energy. Einstein later introduced the notional that the electromagnetic field *itself* was quantized into discrete particles called *photons*, rather than the atoms in the walls, and through this insight explained not only Planck's result but also the photoelectric effect. Both men were awarded the Nobel Prize in physics as a result of their work in this field.

The Planck-Einstein relation is most often stated as:

$$E = hf \tag{1.14}$$

where h is the *Planck constant*: $6.62607004 \times 10^{-34} \text{m}^2\text{kg/s}$. In this book, we'll often use *angular frequency* ω (phase angle per unit time), rather than *linear frequency* f (complete cycles per unit time) as a measure of frequency, so it will be convenient to introduce $\hbar = h/(2\pi)$ so that the relationship can also be written as:

$$E = hf = \frac{h}{2\pi}\frac{2\pi}{T} = \hbar\omega \tag{1.15}$$

As we study quantum mechanics we're going to apply this relationship not just to photons, but to *many* quantum behaviors. Exactly how this happens will become clear as we go, but keep it in mind as we continue our study of quantum mechanics.

1.7 THE DE BROGLIE HYPOTHESIS

In his 1924 Ph.D. thesis Louis de Broglie postulated that all particles of matter (not just photons) have wave properties. Einstein had introduced the idea of quantized electromagnetic waves (now called *photons*). De Broglie supposed that the momentum of a particle could be connected to a *wavelength* associated with that particle through the same relationship used for photons, namely:

$$p = \frac{h}{\lambda} \tag{1.16}$$

where h is the Planck constant and λ is the *wavelength* of the particle. This was a *revolutionary* idea. Also, just as we introduced the *angular* frequency, we'll often prefer to use wave-number: $k = (2\pi)/\lambda$ (phase per unit distance), rather than wavelength to describe the spatial variation of a quantum wave:

$$p = \frac{h}{\lambda} = \frac{h}{2\pi}\frac{2\pi}{\lambda} = \hbar k \tag{1.17}$$

1.8 THE HEISENBERG UNCERTAINTY PRINCIPLE

One of Werner Karl Heisenberg's most famous insights is now known as the "Heisenberg Uncertainty Principle". The idea is that improved knowledge of one observable property of a system may require that knowledge of another property is diminished, in such a way that the *product* of the two uncertainties always remains greater than a minimum value dictated by nature. This is often written as:

$$\Delta x \Delta p \geq \frac{\hbar}{2} \tag{1.18}$$

As we learn more quantum mechanics we'll see how this result is a consequence of the way superpositions of quantum mechanical wavefunctions behave, and how observable properties are determined through quantum *operators* acting on those wavefunctions. For now it's enough to simply remember that sometimes quantum mechanics forbids us from knowing certain related observable properties of a system simultaneously.

1.9   PRACTICAL CALCULATIONS

Let's end this chapter with some practical examples. It's helpful to reduce abstraction by focusing our attention on concrete situations. Remember that the universe is made up of real particles with concrete properties that are known. Quantum mechanics is often perceived as highly abstract and theoretical. In order to peel away the layers of theory and abstraction, we can "bring it down to Earth" with some concrete examples. In the next chapter we'll explore the most basic quantum mechanical example of all: a particle confined to a 1D box. We'll learn that there are only a specific set of values of wavelength (λ) allowed for such a particle to fit into the box. So, if the box has a length of L the allowed wavelengths have to be integer fractions of $2L$ (we'll get into *why* this is so in the next chapter). But just having this information is enough to compute the allowed energies of the particle. Amazing! As a practical matter it is useful to keep some numbers handy in our working memory. One is the mass energy of some common particles (tabulated here in Table 1.1).

TABLE 1.1   Approximate Mass Energies of Particles

Particle	Mass Energy
Electron	0.511 MeV
Muon	106 MeV
Proton	940 MeV
Neutron	940 MeV

Another is the product of $\hbar$ and c (the speed of light). This can be expressed in handy units for practical calculations as:

$$\hbar c = 197 \text{eV.nm} = 197 \text{MeV.Fm} \tag{1.19}$$

How is this useful? Let's see! Let's say I have an electron ($mc^2 = 0.511\text{MeV}$) confined to a 1D box of length 3 nm. Suppose we want to compute the lowest possible energy such an electron could have according to quantum mechanics. For reasons we'll learn in the next chapter the longest possible wavelength such an electron could have is $2L$. But this means the lowest possible value of k would have to be:

$$k = \frac{2\pi}{\lambda} = \frac{2\pi}{2L} = \frac{2\pi}{2}\frac{1}{L} = \frac{\pi}{L} \tag{1.20}$$

But recall from the de Broglie relationship that k corresponds to momentum: $p = \hbar k$, so that means we have momentum with a magnitude of:

$$p_n = \hbar k = \frac{\hbar \pi}{L} \tag{1.21}$$

Assuming the potential energy in the box is zero, the only type of energy that contributes to the total energy is kinetic energy:

$$E = KE = \frac{p^2}{2m} = \frac{(\hbar k)^2}{2m} = \frac{(\hbar \pi)^2}{2mL^2} = \frac{\hbar^2 \pi^2}{2mL^2} \tag{1.22}$$

That's it! But wait, what's the value? You can naturally just plug in numbers and compute E, but wouldn't it be nice to estimate quickly without having to get out a calculator?

Sure it would! But how? Easy, use the known value of $\hbar c$ and the known value of the mass-energy of the electron from Table 1.1 to substitute approximate values. Just multiply m by c^2 and $\hbar$ by c:

$$E \quad = \quad KE = \frac{p^2}{2m} = \frac{(\hbar k)^2}{2m} = \frac{(\hbar \pi)^2}{2mL^2} = \frac{\hbar^2 \pi^2}{2mL^2} \tag{1.23}$$

$$= \quad \frac{(\hbar c)^2 \pi^2}{2(mc^2)L^2} = \frac{(197\text{eV.nm})^2 \pi^2}{2(0.511 \times 10^6 \text{eV})L^2} \tag{1.24}$$

$$\approx \quad \frac{(200\text{eV.nm})^2 \cancel{\pi^2}}{\cancel{2}(\cancel{0.511} \times 10^6 \text{eV})\cancel{(3\text{ nm})^2}} \approx \frac{4 \times 10^4 \text{eV}}{10^6} \approx 0.04\text{eV} \tag{1.25}$$

We got a bit lucky with the numerical coincidence that $3\,\text{nm} \approx \pi$ and $2 \times 0.511 \approx 1$ but the point is that you can use these numbers to make quick estimates without having to pull out the computer or the calculator. By the way, the more exact answer is 0.0416 eV so it's within about 5% of our estimate. Not bad for the back of an envelope!

1.10   REPRESENTING A PHASOR WITH VPYTHON

We'll be using the VPython package to represent various wavefunctions. Each wavefunction will be represented by a collection of *phasors* each of which is an arrow object in VPython. An arrow has three main features: a position, an axis, and a shaftwidth. The position is the location of the tail of the arrow, the axis is a vector that goes from the tail to the tip, and the shaftwidth determines how wide the shaft of the arrow is. As an example, you can create an arrow that points from the origin to the point (1,2,0) with a shaftwidth of 0.1 using very short program like this:

```
import vpython as vp

myArrow = vp.arrow(pos=vp.vector(0,0,0), axis=vp.vector(1,2,0),
    shaftwidth=0.1)
```

which produces a nice 3D arrow similar to the one displayed in Fig. 1.4.

How can we use this to create a 3D representation of a phasor? The most natural way would be to have a function that accepts an arrow and a complex number. It would modify the arrow object to represent the complex number as a phasor, something like this:

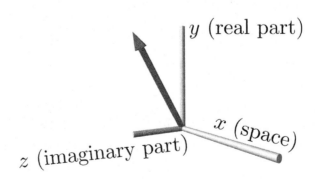

Figure 1.4   A single 3D phasor rendered in VPython.

```
def setArrowFromComplexNumber(a, psi):
    """
    Modify the arrow object, 'a' to represent the complex number 'psi'
        as a phasor.
    """
    a.axis.x = 0
    a.axis.y = psi.real
    a.axis.z = psi.imag
```

Why use the y and z directions? As we'll learn in Chapter 2 we're going to be using the x direction to represent real physical space, and the phasors we'll be drawing will be representing complex numbers that vary with position (x) and time. The real and imaginary parts of those complex numbers will be represented by phasors that point into the y-z plane. Now we can use this function, along with a phasor from Example 1.2 to represent that phasor as a 3D object. Note that we'll be using VPython `cylinder` objects to represent the coordinate axes.

```
import vpython as vp
import numpy as np

def setArrowFromComplexNumber(a, psi):
    """
    Modify the arrow object, 'a' to represent the complex number 'psi' as a
        phasor.
    """
    a.axis.x = 0
    a.axis.y = psi.real
    a.axis.z = psi.imag

vp.scene.background = vp.color.white

# represent the x, y, and z axes using cylinders aligned with those
    directions

vp.cylinder(axis=2*vp.vec(1,0,0), radius=0.05)
vp.cylinder(axis=2*vp.vec(0,1,0), radius=0.05)
vp.cylinder(axis=1*vp.vec(0,0,1), radius=0.05)

myArrow = vp.arrow(pos=vp.vector(0,0,0), shaftwidth=0.1, color=vp.color.red)

# recreate the phasor from example 1.2

R = 2
theta = np.pi/6
z = R*np.exp(1j*theta)

# now render that phasor using 'myArrow'

setArrowFromComplexNumber(myArrow, z)
```

We'll see in Chapter 2 how this approach can be extended to represent arbitrary 1D complex wavefunctions using 3D phasors.

1.10.1 Exercise

Rerun the program with different values of R and θ. Also, print out the actual value of z. What happens when $\theta = \pi/2$? Does that make sense? How about $\theta = \pi$? Is that consistent with what you see in the 3-D view?

Visualizing 1D Quantum Wavefunctions

ONE OF the most basic ideas of quantum mechanics is that the condition of a system is represented mathematically by a *wavefunction*. This is a complex valued mathematical function of that tells us everything we can know about the system. In order to make predictions we apply mathematical operations to the wavefunction that tell us the probability of various outcomes that we can compare to experimental results. The wavefunction is a *complex* valued function. In other words when you evaluate it for a particular set of inputs (e.g., position, time, momentum, energy, etc.) the output is a complex number. The *magnitude* of the complex number gives information about the relative likelihood of a particular outcome. That seems pretty odd I'm sure. The best way to get comfortable with the notion is to play around with it. Let's consider an example.

2.1 EXAMPLE: A PARTICLE AT REST

The simplest example of a "quantum mechanical system" is a single particle at rest.[1] Of course if we *know* it's at rest then $\vec{p} = 0$ and, by Heisenberg's principle of uncertainty relating the uncertainty in particle's momentum to the uncertainty in its position, it follows that we must have no idea *where* the particle is. What's the simplest wavefunction that could be employed to represent this kind of situation? How about a simple constant, like Eq. 2.1?

$$\Psi(x,t) = A \tag{2.1}$$

This seems super simple. However, if the particle has energy we know the Einstein relation requires that it have a frequency proportional to the energy, so we need to add that as a time-dependent phase:

$$\Psi(x,t) = Ae^{-i\omega t} \tag{2.2}$$

where ω is related to the energy by the Plank-Einstein relation: $\omega = E/\hbar$.

---

[1]Feynman, R. P., Leighton, R. B., & Sands, M. L. (1965). The Feynman Lectures on Physics: Quantum Mechanics (Vol. 3). Reading, MA: Addison-Wesley.

DOI: 10.1201/9781003437703-2

2.1.1 Visualizing in Space-Time, and in 3D

We can begin to visualize this kind of wavefunction by drawing lines of constant *phase* on a space-time diagram like that shown in Fig. 2.1. A space-time diagram shows events at different times (the vertical direction, measured in this example in seconds) and also at different positions (the horizontal direction, measured in "light seconds" abbreviated as "csec"). The nice thing about a space-time diagram is that you can easily determine the measurements made by observers in one frame of reference based on the relative time and positions of events in some other frame of reference. Let's set the stage in our own frame of reference with the stationary particle. In this diagram, the dashed and dotted lines represent lines of constant phase. The period of this particular wavefunction is 0.5 sec., the frequency is 2.0 Hz, so the energy works out to be around $E = \hbar\omega \approx 8 \times 10^{-15}$eV. This is an exceedingly tiny energy! In most of the practical examples we'll consider the frequencies (and energies) will be much higher, but 2 Hz is easy to contemplate to begin with. Note that since the lines of constant phase are horizontal in this diagram it means that at any particular time the phase will be the same at all locations in space. This is essentially what Eq. 2.2 is saying. Also, note that events A, B, and C are examples of events that occur at the same place (A and B both occur at $x = 0.0$ csec) and at the same time (B and C occur at $t = 0.5$ sec). An observer in this frame would experience a single event (A) at $t = 0$ sec and then two simultaneous events at different locations (B and C) at $t = 0.5$ sec.

What would this look like in a phasor representation in 3D? Since at any moment in time the phase of the wavefunction is the same at every position then the 3D picture of the phasors at any time will be a collection of phasors that all share the same phase, illustrated in Fig. 2.2. Note that the three phases shown in Fig. 2.2 are the same as the solid, dashed, and dotted lines in Fig. 2.1. Each of the three phases shown are separated by the same phase difference of $2\pi/3$ or one-third of a complete oscillation.

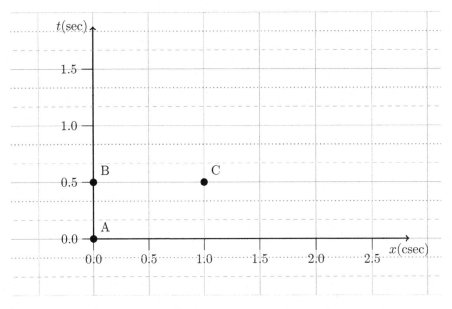

Figure 2.1 A space-time view of a complex wavefunction of a stationary particle with zero momentum and uncertain position. The solid, dashed, and dotted lines are lines of constant phase $(0, 2\pi/3, 4\pi/3$, respectively). The period and frequency are 0.5 sec and 2 Hz, respectively. Note that events A, B, and C are examples that occur at the same place (A and B) and at the same time (B and C).

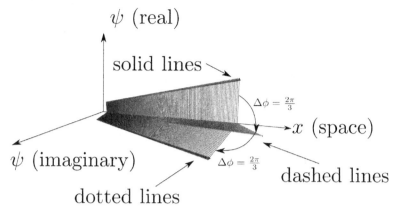

Figure 2.2 The 3D representation of a 1D Complex Wavefunction at a series of three successive times. The phase of the wavefunction varies in time as indicated by the rotation of the phasors in the complex plane. Note that the three phases shown correspond to the solid, dashed, and dotted lines in Fig. 2.1. Each of the three phases shown are separated by the same phase difference of $2\pi/3$ or one-third of a complete oscillation.

2.1.2 Visualizing from a Moving Observer's Perspective

Next let's consider how this situation would appear to someone viewing the particle's wavefunction from a moving frame of reference. Suppose we have an observer moving to the right at half the speed of light. We can label the space-time axes for this observer x' and t'. As you can see in Fig. 2.3 the observer in this moving (primed) frame of reference will think

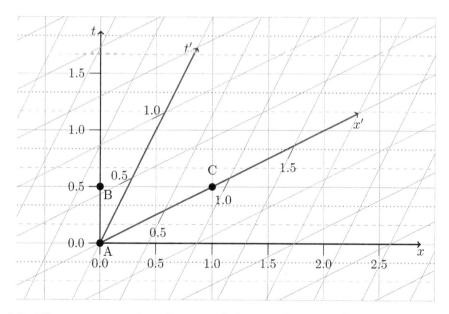

Figure 2.3 The same space-time diagram of the wavefunction of a particle stationary in the unprimed frame of reference with events A, B, and C. Superposed on this frame is the primed frame of a moving observer. In the primed frame, an observer would see a different phase behavior and the three events would occur in a different order.

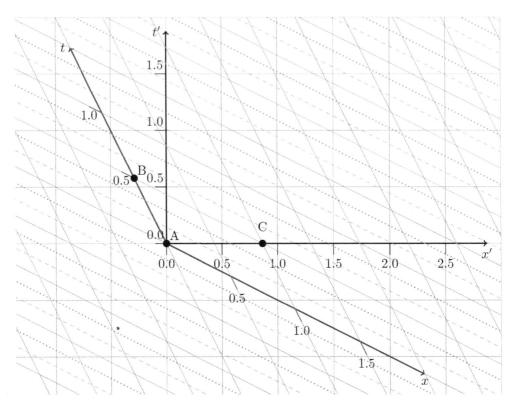

Figure 2.4 The same space-time diagram of the wavefunction of a particle stationary in the unprimed frame of reference with events A, B, and C. This version is drawn with the primed axes in the vertical and horizontal orientation. It's more clearly illustrated here that in the primed frame, an observer would see a different phase behavior and the three events would occur in a different order.

differently from the observer in the unprimed frame where the particle is at rest. First, notice that the events A, B, and C now occur in a different sequence. Lines of constant t' are not horizontal as in the unprimed frame, but they slope up and to the right. In the primed frame the events B and C are no longer simultaneous. The primed observer will experience events A and C as simultaneous events first, and then, more than 0.5 sec. Later, event B will occur. Notice also that the dashed and dotted lines in Fig. 2.3 are not parallel to the x' axis. What is the consequence of this? It means that the phase of the wavefunction will vary in space at any moment in time for the observer in the primed frame. How can we visualize this more clearly? What if we transform the space-time diagram so that the primed observer's axes are vertical and horizontal? This is illustrated in Fig. 2.4. In this view it's clear that the events A and C are now simultaneous and event B now occurs at a different place (to the left of the origin) and at a much later time (in fact it occurs *after* 0.5 sec, even later than it did in the unprimed frame). Of course the goal here is not to learn relativity theory, but to understand how the wavefunction of a particle can be viewed from a moving frame of reference. If you're interested in learning more about space-time diagrams, see the recommended background readings cited in the Introduction. The good news is that we'll not be needing a lot of relativity, but it's worthwhile to understand that the connection between the wavelength of a particle with well-known momentum and it's motion is actually a relativistic effect. How's that? Let's look at the 3D view of the wavefunction from the point

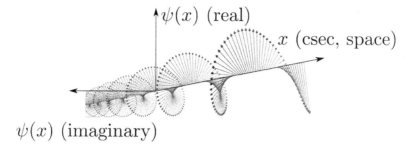

Figure 2.5 The 3D representation of the same 1D Complex Wavefunction at a moment in time $(t = 0)$ from the point of view of an observer in the primed frame of reference described in Fig. 2.4. The phase of the wavefunction varies in space at this instant as indicated by the rotation of the phasors in the complex plane.

of view of this observer in the primed frame. The 3D view of the wavefunction is shown in Fig. 2.5. What do we see? Now the wavefunction which, at any moment in time, used to be a bland line of phasors all pointing in the same direction now is a rich helical spirally thing with a clear period in both space and time. It's got a wavelength. Out of the simple case of a stationary particle, just looking at it from the point of view of a moving observer we get that the wavefunction has a wavelength directly related to the observer's motion. Of course, the observer in the primed frame doesn't *think* they're moving, but really, from their point of view, it's the *particle* that's moving. So we see that from this argument we can conclude that a moving particle with a well-known momentum has a wavefunction like that shown in Fig. 2.2.

While the full mathematical derivation of the result is beyond the scope of this book the final result is actually quite simple. It's just the de Broglie result: Eq. 1.17. The wavelength of a wavefunction is related to the momentum of the particle by the de Broglie formula. Also, it's useful to point out that you can *infer* from this that there is a simple mathematical operation that can effectively *extract* the momentum and the energy from the wavefunction itself. This is a key insight for interpreting the Schrödinger Wave Equation (SWE). You'll note that in a frame of reference where the particle is moving we end up with a wavefunction Eq. 2.3.

$$\Psi(x,t) = Ae^{i(kx-\omega t)} \tag{2.3}$$

We also know that the momentum of such a particle is $\hbar k$, and the energy is $\hbar\omega$. Note that if you were to take the partial derivative of $\Psi(x,t)$ with respect to x, you'd get $ik\Psi(x,t)$. If you multiply this by $-i\hbar$ you'd get the momentum *times* $\Psi(x,t)$. This is one way to think of the *momentum operator* $-i\hbar\partial/\partial x$. It is the mathematical operation you'd perform on $\Psi(x,t)$ to get the momentum *times* $\Psi(x,t)$. Likewise if you take the derivative of $\Psi(x,t)$ with respect to time you'd get $-i\omega$ *times* $\Psi(x,t)$. If you multiply this by $i\hbar$ you'd get the energy *times* $\Psi(x,t)$. In this sense the energy operator can be thought of as $i\hbar\partial/\partial t$. Note that from these operators we can also build up other operators. For example, since KE $= p^2/2m$ we can use the momentum operator to get the kinetic energy operator KE $= -\hbar^2(\partial^2\Psi(x,t)/\partial x^2)/(2m)$. This notion of momentum and energy operators also gives us a simple way to interpret one of the most important equations in quantum physics: the SWE. The SWE is a statement about energy. It says essentially that if you apply the KE operator plus the PE operator to a wavefunction you should get the same result you'd get if you apply the E operator to the same wavefunction. In other words:

$$\cdot (KE + PE)\Psi(x,t) = -\frac{\hbar^2}{2m}\frac{\partial^2}{\partial x^2}\Psi(x,t) + V(x)\Psi(x,t) = i\hbar\frac{\partial}{\partial t}\Psi(x,t) = E\Psi(x,t) \quad (2.4)$$

While we justified this statement using simple plane wave solutions of free particles, it turns out it's much more fundamental. This result can be used to find wavefunctions in much more general situations, as we'll see.

2.2 REPRESENTING A WAVEFUNCTION WITH VPYTHON

How can we represent a simple wavefunction in VPython? Let's extend our work from Chapter 1 to use a collection of arrows to represent the stationary particle in the unprimed frame of reference. We can start out with the same setup we used in Chapter 1 to render a single phasor:

```
import vpython as vp
import numpy as np

def setArrowFromComplexNumber(a, psi):
    """
    Modify the arrow object, 'a' to represent the complex number 'psi' as a
        phasor.
    """
    a.axis.x = 0
    a.axis.y = psi.real
    a.axis.z = psi.imag

vp.scene.background = vp.color.white
```

Next we'll need to create a collection of arrows to represent the value of the wavefunction at different positions in space as 3D objects. To manage a collection of arrows, we'll use the python list object. We'll also use a numpy array to evaluate the wavefunction itself. If you're not familiar with the concept of a python list, or the features of the numpy library, please consult Appendix B for a short introduction.

```
N = 40 # how many arrows?
L = 20 # over what range of x values do we want to span?

x = np.linspace(-L/2,L/2,N)  # generate the x values as a numpy array
arrows = []  # start with an empty python list to hold the arrows

for xval in x: # iterate through the x values in "x"
    # create a red arrow at each xval
    a = vp.arrow(pos=vec(xval,0,0), axis=vec(0,1,0), color=vp.color.red)
    arrows.append(a) # add each arrow to the list as they are created
```

Now we have a collection of arrows in the **arrows** list. Any time we need to update the wavefunction, we can simply loop over the arrows and update each one to represent the value of the wavefunction at the corresponding value of x. For example, let's define a python function `psi(x)` to evaluate the un-normalized wavefunction for the case when the momentum and energy are both well known. As we've seen this is a plane-wave solution $\psi(x) = Ae^{i(kx-\omega t)}$. What should we do to compute the normalization constant A? For now, let's leave this since it just scales the wavefunction but doesn't affect the *shape*. We'll see how this shows up, and makes more sense, in more realistic situations later. For the moment, we'll just use a default value of 1. As long as we don't need to answer any questions about the probability of finding the particle in some range of x values, this value is arbitrary anyway. Let's write a python function that evaluates this function:

```
T = 0.5
omega = 2*np.pi/T
k = 0  # particle is stationary, so k=0
A = 1

def psi(x, t):
    """
    Evaluate the un-normalized wavefunction at the point 'x' and time 't'.
    """

    return A*np.exp(+1j*k*x - 1j*omega*t)  # un-normalized wavefunction, A=1
```

So now the plan is clear. If we need to display the wavefunction at any particular time, we can simply call this function to first *evaluate* the wavefunction, and then loop over the arrows and update their direction and magnitude using the `setArrowFromComplexNumber` function. We'll be using this same plan in virtually every program we write. This is a core concept of visualizing quantum wavefunctions with VPython.

```
t = 0
psiArray = psi(x, t) # evaluate the wavefunction

for i in range(N):
    setArrowFromComplexNumber(arrows[i], psiArray[i]) # update the arrows
```

When you run this code you'll see a visualization of the wavefunction at $t = 0$ that looks like Fig. 2.6. You can use the QR code to see the program in 3D.

How can we turn on the time? Easy. Just iterate this process with ever increasing values of t. See the code below. Note that this code will animate the time evolution for 5 sec, with a time step of 0.1 sec per step.

Figure 2.6   The 3D representation of a particle at rest at time $t = 0$.

```
t = 0
dt = 0.02

while t<5: # simulate for 5 seconds
    vp.rate(10)  # animate at 10 frames per second
    psiArray = psi(x, t) # evaluate the wavefunction

    for i in range(N):
      setArrowFromComplexNumber(arrows[i], psiArray[i])  # update the arrows

    t = t + dt # increment the time and move on
```

Note that we used a new **VPython** function here: **rate**. The **rate** function is useful for animations that need to modify 3D object properties over time. Without rate all these modifications would be made a very high speeds and would be impossible to view by humans. We use **rate** to specify a maximum frame rate for animation. In this case we want to animate at 10 frames per second which will produce a *real time* animation since the time-step size is 0.1 sec. This is a little chunky, so you may be happier with a time-step of 0.05 or 0.02, but realize that then you'll be viewing the scene in *slow-motion*. If you wanted higher time resolution, you could dial down the time-step even more and increase the frame rate accordingly. When you run this code you'll see all the phasors pointing in the y-z plane in the same direction, but rotating at a constant speed around the x axis. This is exactly what you'd expect.

If you increase k to 1 you'll see the effect of a non-zero momentum on the wavefunction as shown in Fig. 2.7.

Figure 2.7   The 3D representation of a moving particle.

2.3   EXERCISE: 3D WAVEFUNCTION VISUALIZATION

How does changing the value of k affect the resulting wavefunction? What would it mean for k to be negative? Make this change and observe the behavior of the resulting wavefunction. Is it moving in the way you'd expect? How should the wavefunction propagate based on Fig. 2.5? Experiment with different values of k and ω to learn how they affect the wave and its time evolution.

2.4   MORE GENERAL CASES

As we move on to more general situations the wavefunctions will get more complicated, but some things will remain the same. First of all, any general wavefunction can be constructed by adding together wavefunctions with definite energies, and therefore definite *frequencies*.

These wavefunctions are sometimes called *energy eigenfunctions* because they are solutions to an *eigenvalue* problem looking for functions that correspond to systems with well-defined energy. These wavefunctions will always have very simple time evolution that consists of multiplying the spatial part of the wavefunction $\psi(x)$, by a time-dependent phase, just like Eq. 2.2. In a 3D visualization, like Fig. 2.2 this just means the phasors will all rotate at a uniform rate as a group around the x axis.

## CHAPTER 3

Visualizing 1D Bound Systems

MANY INTERESTING systems are "bound" in the sense that the particle is constrained to a finite region by forces that prevent it from escaping. It's easy to visualize this in one dimension by graphing the potential energy as a function of position. In a bound system, the potential energy has at least one minimum somewhere and rises to a maximum value, possibly infinity, on both the right and left. A typical case is illustrated in Fig. 3.1. The potential energy (solid) is low near the original, and rises to the left and right. The sum of kinetic and potential energy is also plotted here (dashed) which is often referred to as the "energy". Note that if you look far enough to the right or left the potential energy (solid) is greater than the energy (dashed). This is the hallmark of a bound system! If this were a *classical* system the particle would be bound between the turning points (-7.5 and $+7.5$ on this graph) since at these points the kinetic energy (the difference between the blue line and the red line) would approach zero. This is where the classical particle would slow down, momentarily stop, and then return to the region of lower potential energy.

In this chapter, we'll consider three examples of bound systems: the Infinite Square Well, the Simple Harmonic Oscillator, and the Finite Square Well. Each of these has some unique properties and behaviors, but they also have many similarities. Then, at the end, we'll touch on the connection between bound systems and scattering. Let's start with the simplest bound system, the Infinite Square Well (ISW).

3.1   THE INFINITE SQUARE WELL

The Infinite Square Well (ISW) is an idealization of a physical system that strictly prohibits the particle from straying beyond a certain range of x values, but exerts no influence over the particle within those limits. So the particle is "free" to roam, but only over a prescribed range on the x axis. An example of this kind of potential energy (PE) function is represented in Fig. 3.2. Notice that the potential energy is *constant* (arbitrarily set to zero, but any constant value would lead to the same behavior) when x is between 0 and 5 (arbitrary distance units) and *infinite* outside that range. We normally refer to the size of the well as the "well width". Suppose we call that width by a variable name, say L. Then we'd say in this case $L = 5$. This means that the particle must definitely be somewhere between 0 and 5 at all times, but within that limited range, the potential doesn't exert any influence that favors any one region over another.

We already found the kind of wavefunction that we'd expect for a *completely* free particle. Specifically, it would need to look like Eq. 2.3. The problem with this solution is that it

DOI: 10.1201/9781003437703-3

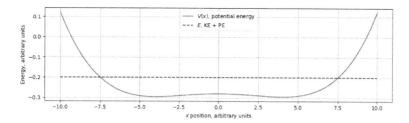

Figure 3.1 Typical potential energy function (solid) with energy (dotted) somewhere between minimum and maximum values of the potential energy.

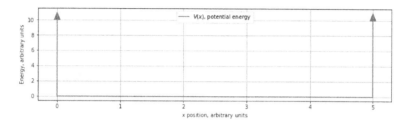

Figure 3.2 The Infinite Square Well (ISW) potential.

cannot, by itself, represent a particle that is constrained to a specific range of x values. As we saw in Chapter 2, the wavefunction of a completely free particle goes on forever! However, we *can* use the superposition of *two* of these solutions to get a solution that can work. Before we discuss what that means we need to consider the constraints that must be placed on a physically realizable wavefunction. We know the wavefunction must be zero outside the ISW since the potential energy there is infinite. This means there's an infinite force keeping the particle from escaping the limits of the well. This also implies that the wavefunction has to go to zero as it approaches the edges of the well. As we'll see in a moment, this requirement limits the values of k and ω we can use in any superposition of free particle wavefunctions.

3.1.1 Continuity

Why does the solution need to go to zero? It turns out that to be physically plausible, all quantum wavefunctions need to be finite and *continuous* functions of position. Since the potential is infinite outside, you'll recall that we used the free particle wavefunction to infer the meaning of the momentum $(-i\hbar\partial/\partial x)$ and energy $(i\hbar\partial/\partial t)$ operators. Applying this interpretation to the current example, the kinetic energy (KE) would be proportional to the second derivative of the wavefunction: $-\hbar^2(\partial^2\Psi(x,t)/\partial x^2)/(2m)$ as described in Eq. 2.4. This has implications for the continuity of the wavefunction and its derivatives. In this example, the potential energy goes to infinity outside the boundaries of the well. This means the KE can become infinite where the PE is infinite so long as it can be done in a way that keeps the sum KE+PE finite. However, for the KE to be *defined* it means the wavefunction has to be *continuous*, otherwise the first derivative would have an infinite value, and the second derivative (the KE) wouldn't even be *defined*. So, we can add multiple solutions together to produce a solution that works for the ISW, but it still needs to satisfy the requirement that the wavefunction is continuous. Since the wavefunction needs to be zero *outside* the boundaries of the ISW, it also needs to be zero *at* the boundaries of the well.

3.1.2 Superposition of Free Particle Solutions

The idea is that we can add two instances of the free particle solution of the SWE (Eq. 2.4) together to form a solution that not only solves the SWE but also satisfies the boundary condition at the edges of the well. For each solution we need to choose the ω and k values carefully so that when we add the solutions together, we can arrange things so the superposition goes to zero at the edges of the well. Then we can simply *use* that superposition within the well and let the wavefunction be zero outside the well. Let's try two solutions with the same energy. Since the energy for a free particle is only kinetic and KE $= p^2/2m$, but that's $(\hbar k)^2/2m$. You can see there are two distinct values of k that have the same energy: $k = +2\pi/\lambda$ and $k = -2\pi/\lambda$. Let's add those two together and see what we get:

$$\Psi(x,t) = A_+ e^{i(+kx-\omega t)} + A_- e^{i(-kx-\omega t)} = A\sin(kx)e^{-i\omega t} \tag{3.1}$$

Here A_+, A_- are just arbitrary constants that we can determine based on the need for the total probability of finding the particle anywhere to be 1, and the need to satisfy the boundary conditions. By choosing A_+ and A_- to be $\pm\frac{A}{2i}$ respectively we get the result, $A\sin(kx)$ for the x-dependent part of the wavefunction. This has the nice effect of satisfying the boundary condition on the left side of the well automatically since $\sin(kx)$ will always be zero there. We still need to pick a value for the overall amplitude A. That process is called *normalization* and we'll get to that shortly. However, the right side of the well is another matter. For that to work, we need $\sin(kL) = 0$ so when $x = L$ the wavefunction goes to zero. But $\sin(kL)$ is zero only when $kL = n\pi$, so this means the values of k that work are *quantized*. This is how quantization arises! In order for the boundary conditions to be satisfied we need $k = n\pi/L$ where n is a natural number (e.g., 1, 2, 3, etc.). Note also that by Eq. 2.4 when k changes, so must ω since $E = p^2/2m$ it means $\hbar\omega = (\hbar k)^2/2m = (\hbar n\pi/L)^2/2m$ so $\omega = \hbar(n\pi)^2/(2mL^2)$, or more simply $\omega = n^2\omega_1$ where ω_1 is the angular frequency of the ground ($n = 1$) state. These states with a single value of ω are called energy *eigenstates*. They have a well-defined value of energy. They are also called *stationary* states since, as we'll soon see, they have probability distributions that don't change over time.

3.1.3 Visualizing ISW Quantum States

How can we visualize these solutions? Easy! We can build on the solutions from the previous chapter. We only need to change the `psi` function and the coordinates of the arrows. You can start with the program for the free particle and adjust the x values:

```
x = np.linspace(0,L,N)  # generate the x values as an array
vp.scene.center = vp.vec(L/2,0,0)
```

Since the coordinate system for the ISW goes from 0 to L rather than $-L/2$ to $+L/2$ we need to move the `scene.center` to point to the middle of the well. To visualize the ground state you can set $k = \pi/L$ and change the `psi` function to match:

```
k = np.pi/L  # ground state

def psi(x, t):
    """
    Evaluate the un-normalized wavefunction at the point 'x' and time 't'.
    """

    return np.sin(k*x)*np.exp(-1j*omega*t)  # un-noralized wavefunction, A=1
```

Figure 3.3 Arrows representing the ground state wavefunction of the ISW.

When you run this program you should see arrows that look like this rotating, in phase, about the x axis with a definite frequency (ω). At the end you should see something like Fig. 3.3.

Note that the length of the arrows does not change over time. Since the probability of finding the particle at any location on the x axis is proportional to the squared length of each arrow, it means the probability of finding the particle at any location is not changing with time, even though the phase of the arrows is clearly changing. In this sense these states of fixed ω are sometimes called *stationary* states.

3.1.4 Normalize the ISW Wavefunction

In order to compute probabilities and expectation values from wavefunctions they need to be normalized. For analytical solutions to the SWE normalization is usually done using calculus. We need to demand that the total probability to find the particle anywhere has to be 1 (or 100%). If you apply this requirement to a wavefunction like Eq. 3.1 you'll see that A needs to be $A = \sqrt{2/L}$. When simulating a wavefunction in python it turns out to be much more convenient to think of each arrow representing the amplitude to find the particle in a particular "chunk" of the x axis, and in this way rather than performing a continuous integration over the x axis, we can perform a discrete *sum* over the array of psiArray values. Numpy makes this easy using the sum method, as follows:

```
psiArray = psiArray/np.sqrt((abs(psiArray)**2).sum())
```

Note that exponentiation in python uses the '**' operator rather than '^'. The idea is to scale psiArray by a factor such that when you take the sum of the squares of the absolute values of the psiArray elements you'll get 1. Note that when you do this you may need to scale the arrows in the visualization so that they're still easy to see. One easy way to do this is to find the max arrow length and the scale all the arrows so that they're some nice fraction of the scale L. You can see an example of this in the code below. Also, check with QR code for Fig. 3.4.

3.1.5 Superpositions of Different k Values

What if you have a quantum state that's a superposition of equal parts $n = 1$ and $n = 2$ states? Can we visualize that too? Of course! In this case we need to build psi as a superposition:

```
def psi(x, t):
    """
    Evaluate the un-normalized wavefunction at the point 'x' and time 't'.
    """
```

Figure 3.4   Arrows representing an equal superposition of $n = 1$ and $n = 2$ states.

```
psi1 = np.sin(k*x)*np.exp(-1j*omega*t)     # ground state contribution
psi2 = np.sin(2*k*x)*np.exp(-4j*omega*t)   # first excited state
                                             contribution
sup = psi1 + psi2                          # compute the superposition
return sup/np.sqrt((abs(sup)**2).sum())    # normalize and return
```

Then we'll need to adjust the arrows and timestep to make the interaction between the two states clearer:

```
for xval in x: # iterate through the x values in "x"
    # create an arrow at each xval
    a = vp.arrow(pos=vec(xval,0,0), shaftwidth=0.01*L,
        axis=vec(0,1,0), color=vp.color.red)
    arrows.append(a) # add each arrow to the list


t = 0
dt = 0.001
maxpsi = max(abs(psi(x, 0)))
```

Also in the loop we'll need to rescale the arrows to make the interaction easier to see:

```
while t<5: # simulate for 5 seconds
    rate(100)  # animate at 10 frames per second
    psiArray = psi(x, t)

    for i in range(N):
      setArrowFromComplexNumber(arrows[i], L*psiArray[i]/(2*maxpsi))

    t = t + dt # increment the time and move on
```

After these changes, when you run this program you should see arrows undulating in a way that the amplitudes seem to "slosh" back and forth from left to right as the superposition phase changes over time.

There are a number of features one can notice about this superposition. Since the energy eigenstates have energies that go like n^2, you can see that the $n = 2$ state has four *times* the energy, and therefore four times the *frequency* of the ground state $(n = 1)$. It's also interesting to graph the expectation value of the position as a function of time. How do we do that? Easy! Since the wave function is normalized already we can simply use the probability that the particle's position is associated with each arrow: $|\psi(x)|^2 = $ abs(psi(x,t)**2) multiplied by the position, and then summed over all the arrows. This is easy with numpy. We can just update the main loop like so:

```
while t<0.5: # simulate for 5 seconds
    rate(100)  # animate at 10 frames per second
    psiArray = psi(x, t)

    for i in range(N):
      setArrowFromComplexNumber(arrows[i], L*psiArray[i]/(2*maxpsi))

    xexp = (abs(psiArray)**2*x).sum() # compute <x> using the numpy array
        psiArray
    gr.plot(t,xexp) # plot the expectation value <x>

    t = t + dt # increment the time and move on
```

The gr plotting/graphing object is created at the beginning of the program like so:

```
gd = vp.graph(xtitle="t", ytitle="<x>", width=640, height=300)
gr = vp.gcurve(color=vp.color.black)
```

You can see the full program if you navigate to the QR code from Fig. 3.4.

3.1.5.1 *Exercise: ISW 3D Wavefunction Visualization*

Use the superposition example to graph the expectation value of position as a function of time for the case when there's a varying contribution from the $n = 1$ state and the $n = 2$ states. What happens to the expectation value of x when the superposition is made up of only one state? Does this make sense? Explain.

3.2   THE SIMPLE HARMONIC OSCILLATOR

It's not terribly difficult to apply a similar approach to the one we just took with the ISW to the simple harmonic oscillator (SHO). What's the same? There are still energy eigenstates (stationary states) that have definite energy (frequency). You can still form superpositions of these stationary states to produce more general states that "slosh" in time. These superposition states can be used to model realistic scenarios that might occur in various situations. What's different? Well, the potential energy function is pretty different, rather than a piecewise constant potential like the infinite square well, or the finite square well, the SHO potential is $V(x) = \frac{1}{2}m\omega^2x^2$, and this results in quantitatively different eigenstate energies and wavefunctions.

3.2.1   Qualitative Similarities Between the ISW and the SHO

We'll leave the derivation of the energy eigenstate wavefunctions to the textbooks, however, even though they are *quantitatively* different they are qualitatively quite similar to the ISW states, and that's useful for building intuition! The SHO energy eigenstates, like the ISW eigenstates, are numbered by their eigenvalues ($n = 0, 1, 2, \ldots$ etc.) though for the SHO the numbering starts at $n = 0$ rather than $n = 1$. This is a simple result of the fact that

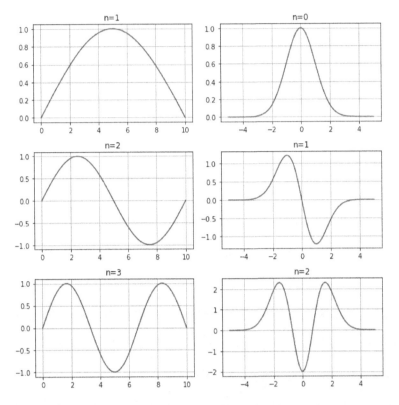

Figure 3.5 Compare ISW (left) and SHO (right) eigenstate wavefunctions.

the $n = 0$ ISW wavefunction would be zero everywhere. The lowest energy available to the ISW is $n = 1$. In contrast the SHO eigenstate wavefunctions are all products of Gaussians multiplied by nth order polynomials (the Hermite polynomials). Since there is nothing pathological about a zeroth order polynomial (i.e., a constant) times a Gaussian, it means that the lowest energy eigenstate of the SHO is $n = 0$. The qualitative similarity between the SHO and ISW eigenstate wavefunctions can be clearly seen in Fig. 3.5. There are a few important features that should be emphasized:

- The general shape of each ISW wavefunction matches very nearly the corresponding SHO wavefunction except that the n values are "off by one". This is the consequence of the ground state being most conveniently defined as $n = 1$ for the ISW but $n = 0$ for the SHO.

- The ISW wavefunctions are strictly constrained by the limits of the well. The SHO wavefunctions "spread out" as n increases.

- The energy of the ISW eigenstates increases as n^2, while the SHO energies are linearly related to n ($E_n = \hbar\omega(n + 1/2)$).

3.2.2 Visualizing SHO Quantum States in VPython

Calculating the SHO energy eigenstates isn't too bad if you take advantage of the recursion relationships between the Hermite polynomials:

```
NA=80                           # how many arrows?
a=15.0                          # range of x is -a/2 to a/2 in units
                                # of $\sqrt{\hbar/m\omega}$
x = np.linspace(-a/2, a/2, NA)    # NA locations from -a/2 to a/2

NHs=20
hs=np.zeros((NHs,NA),np.double)   # the hermite polynomials, an NHs x
                                    NA array
coefs=np.zeros(NHs,np.double)     # the coherent state coefficients, an
                                    NHs x 1 array
psis=np.zeros((NHs,NA), np.double) # the stationary states, an NHs x NA
                                     array

hs[0]=0*x + 1.0            # zeroth Hermite Polynomial, H0
hs[1]=2*x                  # first Hermite Polynomial, H1

#
# Compute the first NHs Hermite Polynomials,
# use recurrence relation to get the rest of the Hn(x)
#
# (see e.g., http://en.wikipedia.org/wiki/Hermite_polynomials#Recursion
    _relation)
#

for n in range(1,NHs-1):
    hs[n+1]=2*x*hs[n] - 2*n*hs[n-1]
```

Once the Hermite polynomials are available, calculating the wavefunctions as products of Gaussian and Hermite polynomials is a matter of keeping track of the normalizing scale factor for each state, along with the Gaussian and Hermite polynomial factors. These can be kept in a list of energy eigenfunctions, called 'psis[]', for convenience.

```
#
# Get the stationary states using the hs array and compute the
# normalization factor in a loop to avoid overflow
#

normFactor = 1.0/pi**0.25
psis[0]=np.exp(-x**2/2)
for i in range(1,NHs):
    normFactor = normFactor/sqrt(2.0*i)
    psis[i]=normFactor*hs[i]*np.exp(-x**2/2)

#
# Now do the sum to compute the initial wavefunction
#
```

Once the stationary states are computed, one can easily construct arbitrary superpositions of these eigenstates, like this one:

```
coefs = [1,1,1]   # equal parts n=(0,1,2) as an example
#
# Now do the sum to compute the initial wavefunction
#

psi=np.zeros(len(x),complex)
for m in range(len(coefs)):
    psi += coefs[m]*psis[m]

#
# Normalize!
#

psi=psi/sqrt((abs(psi)**2).sum())
```

Next one needs to create the arrows to represent the wavefunction:

```
#
# build the arrows. Scale them on the screen by a factor
# of 3 so they look nice.
#

alist = []
for i in range(NA):
    alist.append(vp.arrow(pos=vp.vec(x[i],0,0), color=vp.color.red))
    SetArrowFromCN(3*psi[i],alist[i])
```

Finally, in the same way we've done this before, we can advance the time and plot the expectation value of the position, just as we did for the ISW. The output is illustrated in Fig. 3.6.

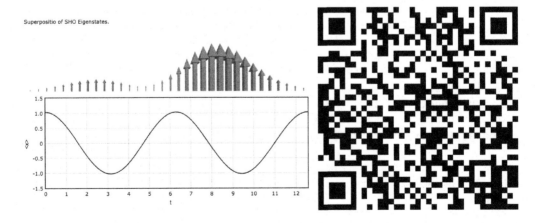

Figure 3.6 Arrows representing the time evolution of an equal superposition state of the SHO. The QR code will lead you to a live version of the program.

```
while t<4*pi:
    rate(1.0/dt)
    psi=np.zeros(len(x),complex)               # start with an empty
                                               wf array

    for m in range(len(coefs)):                # for each basis
                                               function

        psi += coefs[m]*psis[m]*np.exp(-1j*(0.5+m)*t) # add each with time
                                                      dependence

    psi=psi/np.sqrt((abs(psi)**2).sum())       # normalize

    for i in range(NA):
        SetArrowFromCN(3*psi[i], alist[i])     # update arrows, scale
                                               up so they're easier
                                               to see

    pxTot = (x*abs(psi)**2).sum()              # compute the
                                               expectation value of x

    gr.plot(pos=(t,pxTot))  # plot it!
    t += dt
```

Note the similarity to the code to compute the time evolution of the ISW states. This particular superposition state evolves in a way roughly similar to the ISW as shown in Fig. 3.6.

3.2.2.1 Exercise

What is the frequency of oscillation of the expectation value of the position when you include only two neighboring eigenstates? Change the code to verify this. What happens if you only include even eigenstates? Or odd eigenstates? Do you still see oscillations? Compare the magnitude of the oscillation to that when there are neighboring (e.g., $n = 0$, and $n = 1$) eigenstates in the superposition. What's happening? Explain.

3.3 THE FINITE SQUARE WELL

The finite square well has an interesting mixture of the properties of both the ISW and SHO, but also some new, more realistic features, that are present in neither. An example of a FSW is illustrated in Fig. 3.7. Note that *inside* the well, the energy is greater than the potential energy, and the potential energy is constant here. This is exactly the same situation inside the ISW potential, so we shouldn't be shocked if the solution here is also exactly the same! However, unlike the ISW, the potential energy *outside* the well does not go to infinity, so the wavefunction does *not* go to zero at the edge of the well, but instead has a finite value there. The wavefunction *outside* the well only goes to zero in the limit as x goes to $\pm\infty$ away from the well.

At the boundary the wavefunction must be continuous and have a continuous first derivative so that the energy is finite everywhere. This "value and derivative matching" boundary condition turns out to be the constraint that causes the energies of the eigenstates to be quantized, in a way that's similar to the ISW boundary condition determining the quantized energy levels in that case. We don't need to go through the whole derivation here[1] but the

---

[1] Cohen-Tannoudji, *Quantum Mechanics*, p76, Wiley, 2019.

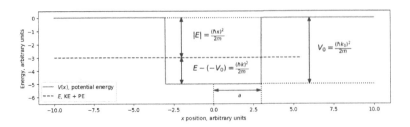

Figure 3.7 A typical FSW potential.

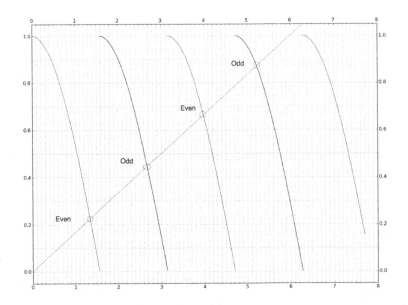

Figure 3.8 Solving for the eigenstates of the FSW. $k_0 a = 6.0$.

result is easy to understand and remember. The depth of the well is V_0 and the "half-width" of the well is a. For any given bound eigenstate energy the energy must be between $-V_0$ and 0. The wavefunctions will look like the ISW *inside* the well, where the energy is greater than the potential energy, but outside the well, where the potential energy is greater than the energy, the wavefunction will be a *real* exponential. The wavenumber inside the well, k is related to $E - (-V_0)$ as shown in Fig. 3.7. However, the wavenumber outside the well, κ, will depend only on the value of the energy E. In order to ensure the wavefunction and its first derivative are continuous one must identify and satisfy one of these two constraints:

$$|\cos(ka)| \; = \; k/k_0; \; (\text{where} \tan(ka) > 0) \tag{3.2}$$
$$|\sin(ka)| \; = \; k/k_0; \; (\text{where} \tan(ka) < 0) \tag{3.3}$$

depending on whether the wavefunction is symmetric (Eq. 3.2) or antisymmetric (Eq. 3.3). These constraints can be solved graphically by finding the values of ka that simultaneously solve the left- and right-hand side of Eq. 3.2 and Eq. 3.3. as shown in Fig. 3.8.

Visualizing these states is similar to the ISW and SHO cases, except that the wavefunctions are made up of piecewise continuous sin(), cos(), and exponential functions. First, using the values of z_1 and z_2 from Fig. 3.8, we can compute values for k_1 and k_2 for the interior of the well, as well as κ_1 and κ_2 in the exterior. These can then be applied to compute the wavefunction solutions inside and outside the well in addition to the frequencies of the two eigenstates, ω_1 and ω_2.

```
z0 = 6.0                        # Not a great choice. Why not? Pick a
                                  better one
k0 = z0/a                       # get k0

#
# numerical solutions for z when z0 = 2.1*pi, you should find a better z0
# and find solutions for that choice.
#

z1 = 1.35
z2 = 2.67

k1 = k0*z1/z0
k2 = k0*z2/z0
kap1 = np.sqrt(k0**2 - k1**2)
kap2 = np.sqrt(k0**2 - k2**2)

E1 = -(hbar*kap1)**2/(2.0*m)
E2 = -(hbar*kap2)**2/(2.0*m)

w1 = E1/hbar
w2 = E2/hbar
wn=[w1,w2]
T = 2*np.pi/(w2-w1) # the approximate period of oscillation

t = 0.0
dt = T/200.0   # a small fraction of a period
```

To actually calculate the wavefunction we need to splice the exterior and interior functions together, matching the value of the wavefunction at the boundary. This is most easily accomplished using the numpy function called piecewise.

```
psis = np.zeros((2,NA),np.double)

def f1(x):
    return np.cos(k1*x)

def f2(x):
    return np.sin(k2*x)

def f3(x):
    return f1(a)*np.exp(-abs(kap1*x))/np.exp(-abs(kap1*a))

def f4(x):
    return np.sign(x)*f2(a)*np.exp(-abs(kap2*x))/np.exp(-abs(kap2*a))
```

```
psis[0] = np.piecewise(x, [x<-a, (x>=-a)&(x<=a), x>a], [f3, f1, f3])
psis[0] = psis[0]/np.sqrt((abs(psis[0])**2).sum())
psis[1] = np.piecewise(x, [x<-a, (x>=-a)&(x<=a), x>a], [f4, f2, f4])
psis[1] = psis[1]/np.sqrt((abs(psis[1])**2).sum())
```

Once the eigenstates are calculated, one can create superpositions and graphs of the position expectation value precisely as was done for the ISW and SHO.

```
# Equal parts 1 and 2
c1 = 1.0/np.sqrt(2)
c2 = 1.0/np.sqrt(2)

cn=[c1, c2]                          # array of amplitudes
t = 0.0                              # start at t=0

psi = np.zeros(NA, complex)                  # construct psi at time '0'
for i in range(len(cn)):
    psi = psi + cn[i]*psis[i]

arrowScale = a/psis[0][NA2]               # scale to make the middle of
                                          # psis[0] about 3a high

def SetArrowFromCN( cn, a):
    """
    SetArrowWithCN takes a complex number  cn  and an arrow object  a .
    It sets the  y  and  z  components of the arrow s axis to the real
    and imaginary parts of the given complex number.

    Just like Computing Project 1, except y and z for real/imag.
    """
    a.axis.y = cn.real
    a.axis.z = cn.imag
    a.axis.x = 0.0

alist = []
for i in range(NA):
    ar = vp.arrow(pos=vp.vec(x[i],0,0), axis=vp.vec(0,a,0),
                  shaftwidth=0.02*a, color=vp.color.red)
    alist.append(ar)
    SetArrowFromCN(arrowScale*psi[i],alist[i])

#
# all the arrows are made, and the basis functions and coefficients are set.
# Create a loop that produces the corresponding time evolution.
#

while t < 2*T:
    rate(30)

    t = t+dt
    psi = np.zeros(NA, complex)
```

```
for i in range(len(cn)):
    psi = psi + cn[i]*psis[i]*np.exp(-1j*wn[i]*t) # a phasor product for
                                                    each eigenstate

for i in range(NA):
    SetArrowFromCN(arrowScale*psi[i],alist[i])

xexp=(x*abs(psi*psi)).sum()
g1.plot(pos=(t,xexp))
```

Again, the superposition states have qualitatively similar behavior as shown in Fig. 3.9.

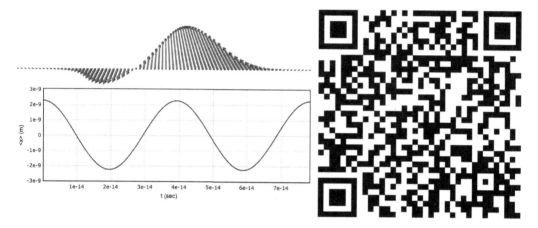

Figure 3.9   Superposition state of the FSW.

3.4   THE WAVE PACKET AND THE FINITE BARRIER

One could object that the "free particle" states we considered in Chapter 2 were not particularly realistic. For one thing, they have infinite extent in both space and time! To create a wavefunction that has a finite extent one can multiply a free particle wavefunction with a precise momentum (e.g., $\psi(x) = Ae^{ikx}$) by a Gaussian envelope (e.g., $e^{-x^2/(2a)}$) function that "clips" the wavefunction in space. The product of these two is a wavefunction whose momentum is "smeared" out around the original momentum value by an amount that depends on the size of the Gaussian envelope. If the Gaussian envelope is large, then the momentum smears only a little, but if the envelope is small, the momentum will smear a lot. This is intuitively clear from the Heisenberg uncertainty principle since the uncertainty in the particle's position is determined by the width of the Gaussian. As the width of the Gaussian grows larger the uncertainty in the momentum decreases. The easiest way to visualize this is to compute and plot the Fourier transform of the product wavefunction as shown in Fig. 3.10.

Note that when the width of the wave packet in real space becomes narrower, the distribution in the Fourier transform (momentum) space becomes broader, and vice versa. This is consistent with the Heisenberg uncertainty principle and is a consequence of the behavior of the Fourier transform between real space and momentum space. What effect does this have on the propagation of the wave packet in space? We'll see that a wave packet has no

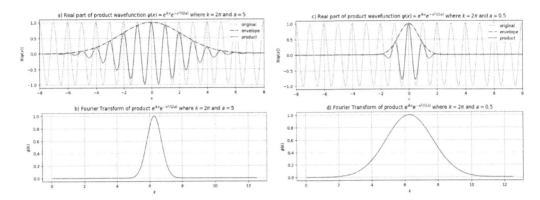

Figure 3.10 Comparing wave packets of different widths in position and momentum space. (a) $k_0 = 2\pi \, \text{rad/m}$, and $a = 5$ m in position space, (b) $k_0 = 2\pi \, \text{rad/m}$, and $a = 5$ m in momentum space, (c) $k_0 = 2\pi \, \text{rad/m}$, and $a = 0.5$ m in position space, and (d) $k_0 = 2\pi \, \text{rad/m}$, and $a = 0.5$ m in momentum space.

choice but to broaden over time. Note that the wave packet must be constructed of momentum components with different wavelengths and speeds. As a result, some components will propagate more slowly, and others will propagate more quickly.

3.4.1 Constructing a Wave Packet in Python

How can we visualize this with Python? We'll use the numpy Fast Fourier Transform `np.fft.fft` module from **numpy** to compute the momentum components of the wavefunction at $t = 0$. For a free particle these momentum components don't change in amplitude over time but only get multiplied by a phase factor, as usual, whose phase is proportional to the energy of each component.

There are some technicalities about the `np.fft.fft` and `np.fft.ifft` functions that we need to understand before we can really use the FFT. The Fourier transform that's returned is really a "discrete" Fourier transform. This means that it's using samples of the wavefunction at discrete values of x, and the function returns an array of complex numbers representing the amplitude of the wavefunction at discrete values of k. Of course, that makes sense since we're representing the wavefunction as an array of complex numbers where each complex number represents the value of the wavefunction at a fixed chunk of the x axis. The lowest non-zero spatial frequency that can be represented in this way is $k_1 = 2\pi/L$, where L is the length of the x axis. The highest spatial frequency that can be represented is $k_{\max} = \pi/\Delta x$ where Δx is the spacing between the x values. The spacing between the k values is $\Delta k = 2\pi/L$, which happens to be the same as k_0. The subtle technical bit is how to interpret the array of complex numbers returned by the `np.fft.fft` function. The convention used by most FFT algorithms, including `np.fft.fft` is that the first N/2 + 1 elements of the array correspond to spatial frequencies from $k = 0$ to $k = k_{\max}$, with a spacing of Δk, and the remaining N/2 - 1 elements of the array correspond to spatial frequencies from $k = -k_{\max}$ to $k = -k_1$. The reason for this is that the FFT algorithm is designed to be used mostly with real-valued functions, and the negative k values are redundant since the Fourier transform of a real-valued function is always symmetric about $k = 0$. However, in quantum mechanics, there is no expectation that the functions being used are real, so we need the entire array of k values. To compute the correct energy for the wavefunction component, we need to understand this technicality and use the correct k values. The code below demonstrates one way to do this.

```
hbar=1.0                        # use units where hbar = 1
m=1.0                           # and m=1.0
NA=500                          # how many arrows?
L=30.0                          # range of x is -L/2 to L/2


x = np.linspace(-L/2, L/2, NA)   # NA locations from -L/2 to L/2
n = np.arange(NA)                # n = array([0,1,2,3,.... N-1])


# this next step computes the correct value of n for the FFT
n = np.piecewise(n, [n<=NA/2, n>NA/2], [lambda n:n, lambda n:n - NA])


k = 2*n*np.pi/L                 # array of correct k values
Energy = (hbar*k)**2/(2.0*m)    # get the kinetic energy array
omega = Energy/hbar             # get the frequency array


t = 0.0
dt = 0.01
kMin = 2*pi/L       # resolution of the k array
k0 = 20*kMin        # pick a value of k0
a = 1               # width of the wave packet in units of L
arrowScale = sqrt(NA*L*2*a)/10.0 # rescale arrows so they're easier to see


psi=np.exp(1j*k0*x - ((x+1*L/4)/(2*a))**2)     # gaussian wave packet
psi = psi/sqrt((abs(psi)**2).sum())            # normalize
phi0 = np.fft.fft(psi)                         # fft at t=0
```

Note that the last step in all this preparation is to actually compute the FFT and store the result in the phi0 array. Each element of this array is the amplitude of one component of the free particle wavefunction. To get the wavefunction at a later time we need to multiply each element of the array by a phase factor that depends on the energy of that component. This is done in the function doStep below. This function also computes the current width of the wave packet, and graphs it, or it computes the current expectation value of the wave packet's position, and graphs that depending on the value of the boolean variable plotSigma.

```
def doStep(plotSigma=False):
    """
    For the current value of "t" compute the wavefunction.
    """

    # multiply each component by the corresponding phase factor
    # then compute the inverse FFT to get the wavefunction at time "t"

    psi=np.fft.ifft(phi0*np.exp(-1j*(omega)*t))

    for i in range(NA):
        SetArrowFromCN(arrowScale*psi[i], alist[i])  # set the arrows
```

```
xexp = (x*abs(psi)**2).sum()
xxexp = (x*x*abs(psi)**2).sum()
sig=sqrt(xxexp-xexp**2)

if plotSigma:
    gr.plot(pos=(t,sig))
else:
    gr.plot(pos=(t,xexp))
```

The 3D visualization of the wave packet is done in the same way as the ISW, SHO, and FSW examples. The only difference is that the wavefunction is computed using the inverse FFT instead of the Hermite polynomials, or the sine and cosine functions. The full program is available at the QR code in Fig. 3.11.

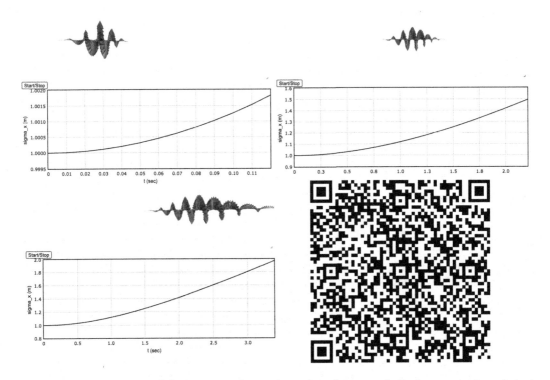

Figure 3.11   The view of the wave packet as it evolves from a relatively compact packet at $t = 0$ to a much broader packet at $t = 3$. The graph shows the value of σ as a function of time.

3.4.2   Scattering from a Finite Barrier

Now that we can construct a wave packet in Python, how do we handle an interaction with a potential? One way is to use the FFT and the inverse FFT to switch back and forth between position and momentum space. We can factor the time evolution $e^{i\hat{H}t/\hbar}$ into two parts: $e^{i\hat{V}t/\hbar}$ and $e^{i\hat{T}t/\hbar}$ where $\hat{V}$ is the potential energy operator, and $\hat{T}$ is the kinetic energy operator. The potential energy operator is diagonal in position space, and the kinetic energy operator is diagonal in momentum space. This means that we can compute the time evolution in

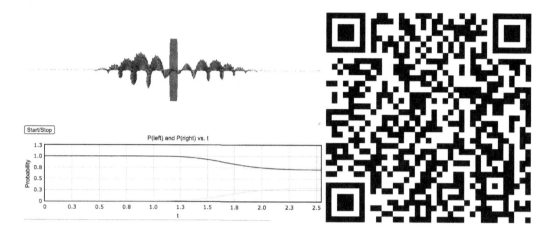

Figure 3.12   A wave packet encounters a finite barrier.

momentum space by multiplying each component of the wavefunction by a phase factor that depends on the kinetic energy of that component, and then we can compute the time evolution in position space by multiplying each component of the wavefunction by a phase factor that depends on the potential energy of that component based on the location of that chunk of the wavefunction.

While it's challenging to explain and probably even more difficult to understand, it's almost trivial to compute! The code below shows how to do it. Note that the potential energy as a function of position is stored in the **V** array, and the kinetic energy divided by $\hbar$ as a function of k is stored in the **omega** array. You can see the 3D output of this program at the QR code in Fig. 3.12.

```
phi=np.fft.fft(psi)
psi=np.fft.ifft(phi*np.exp(-1j*(omega)*dt)) # update free particle state
psi=psi*np.exp(-1j*V*dt)
```

## CHAPTER 4

Visualizing Higher Dimensions

ONCE WE leave the realm of either one dimensional or single particle systems, visualizing quantum wavefunctions becomes more challenging. Suppose, for example, we choose to model a single particle confined to a two-dimensional surface? This is analogous to the 1D systems we encountered in Chapter 3 except that the particle can now move in more than one dimension. The wavefunction then has to depend on both x and y. The trouble is that when we attempt to visualize such a wavefunction we can no longer represent the complex value as a 2D phasor, since after using the x and y direction for the position in space, we only have the z direction left for visualization! Since a phase has two independent parts, either (real, imaginary) or (magnitude, phase). We need some way to visualize these two components distributed over some area. One approach is to use color to represent phase, and height in the z direction to represent magnitude. Another would be to place small objects at a distribution of spatial points and use the properties (e.g., color, size, orientation) to represent various aspects of the wavefunction. In this chapter we'll explore a combination of these approaches and develop some insights from the results.

4.1 THE 2D INFINITE SQUARE WELL

A 2D Infinite Square Well (ISW) is a potential well that has zero potential energy over a finite domain in two directions, say the x and y directions, and is infinite outside that domain. The simplest case is a rectangular domain in the x-y plane with sides L_x and L_y. In this case the 2D TISWE factors into two 1D TISWEs, one in the x direction and one in the y direction. The wavefunction is then a product of the two 1D wavefunctions. The wavefunction is then given by a superposition of energy eigenstate wavefunctions of the form $\psi_{nm}(x,y) = A\sin(n\pi x/L_x)\sin(n\pi y/L_y)$.

These can be visualized using VPython. This section goes through the process of creating a visual representation of a particular solution of a 2D ISW with $L_x = L_y = 1$ and renders the wavefunction for a superposition that starts out with a uniform distribution of probability over the lower left corner ($x < L_x/2$, and $y < L_y/2$) domain. It's similar in spirit to the code from Chapter 3 that was used to visualize the 1D ISW.

The first issue is visualizing the value of the wavefunction at a point in space. In this first example we'll use a vertical cylinder to represent the complex value of the wavefunction. The height of the cylinder will be proportional to the real part of the wavefunction, radius will be proportional to the magnitude of the imaginary part. The color can be set based on the phase of the wavefunction. The code for this is given below.

DOI: 10.1201/9781003437703-4

```
def SetCylinderFromCN( cn, a):
    """
    SetCylinderFromCN takes a complex number  cn  and an arrow object  a .
    This version assumes 'a' is a vp.cylinder and sets the height of
    the vp.cylinder based on the real part, and the radius based
    on the imaginary part. The radius is never set to less than 5% of the
    magnitude of the complex number. The vp.color is set based on the phase.
    """
    a.axis.z = cn.real
    a.radius = max(0.05*abs(cn), abs(cn.imag)/6.0)
    phase = np.arctan2(cn.imag, cn.real)/(2*np.pi)
    a.color = vp.color.hsv_to_rgb(vp.vec(phase,1,1))
```

The next issue is how to represent the wavefunction over a 2D domain. In this case we'll use a grid of cylinders. The code for this is given below. We'll also create a "boundary" to represent the edges of the domain. This is done by creating a set of cylinders that are placed along the edges of the domain.

```
    x, y = np.meshgrid(np.linspace(0,a,NA),np.linspace(0,a,NA))

#
#
# draw the boundaries of the ISW in 2D

boundary_radius = a/100

vp.cylinder(pos=vp.vec(-a/2,-a/2,0), axis=vp.vec(a,0,0),
    color=vp.color.blue, radius=boundary_radius)
vp.cylinder(pos=vp.vec(-a/2,-a/2,0), axis=vp.vec(0,a,0),
    color=vp.color.blue, radius=boundary_radius)
vp.cylinder(pos=vp.vec(a/2,-a/2,0), axis=vp.vec(0,a,0),
    color=vp.color.blue, radius=boundary_radius)
vp.cylinder(pos=vp.vec(-a/2,a/2,0), axis=vp.vec(a,0,0),
    color=vp.color.blue, radius=boundary_radius)


r0 = vector(-a/2, -a/2, 0)    # place origin of arrows

#
# build vp.cylinders.... in 2-D space, store them in a set of nested lists
#

alist = []
for i in range(NA):
    sublist = []
    alist.append(sublist)
    for j in range(NA):
        r = r0 + vector(x[i,j], y[i,j], 0)
        sublist.append(vp.cylinder(pos=r, axis=(0,0,1), color=vp.color.red))
```

Note that the `np.meshgrid` function is a great way to generate x and y coordinates over a 2D grid. The code above uses the same approach as the code in Chapter 3 to generate a 1D array of arrows. The main difference here is that the cylinder positions are now specified in two dimensions.

Next we need to compute the Fourier coefficients, and frequencies for the eigenstates in the superposition.

```
#
# compute the eigenstates and store them in a dictionary 'eigenstates'
#

for nx in NX:
    for ny in NY:
        psinxmy = np.sin(nx*np.pi*x/a)*np.sin(ny*np.pi*y/a)
                                                    # compute the
                                                    # n,m energy
                                                    # eigenstate
        psinxmy = psinxmy/np.sqrt((abs(psinxmy)**2).sum())  # normalize it.
        eigenstates[(nx,ny)] = psinxmy

psi0 = np.zeros((NA,NA),complex)
psi0[:NA2,:NA2] = 1.0
psi0 = psi0/np.sqrt((abs(psi0)**2).sum())        # get psi at t=0, normalized

omega0 = hbar*np.pi**2/(2*m*a**2)                # compute factor omega0

for nmPair in eigenstates.keys():
    nx, ny = nmPair
    psinxmy = eigenstates[nmPair]                     # get nth basis
    cnm = ((psi0[:NA2,:NA2]*psinxmy[:NA2,:NA2]).sum())  # compute fourier
                                                        # coef.
    coefs[nmPair] = cnm                              # save it.
    omega = omega0*(nx**2+ny**2)
                                                    # get omega for
                                                    # nmPair, multiple
                                                    # of omega0
    omegas[nmPair] = omega                          # save it.
```

Next we need to create the 3D scene and set up the initial wavefunction.

```
#
# build up psi via fourier series
#

psi = np.zeros((NA,NA), complex) # initialize psi

for nmPair in eigenstates.keys():
    psi += coefs[nmPair]*eigenstates[nmPair]

for i in range(NA):
    for j in range(NA):
        SetCylinderFromCN(cylinderScale*psi[i,j],alist[i][j])
```

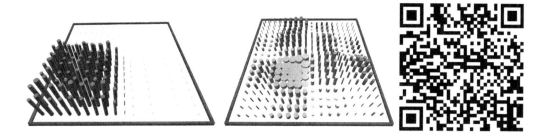

Figure 4.1   Initial wavefunction representation, left. Later time, right, for the 2D ISW.

The variable `cylinderScale` is adjusted to make the cylinders look reasonable.

At $t = 0$ the wavefunction is given by the superposition of the eigenstates that roughly shows the desired behavior. Most of the probability is clearly confined to the lower left-hand corner as expected. At a later time the wavefunction spreads out, similar to the 1D case, but symmetrically across the diagonal. This is shown in Fig. 4.1.

And finally we need to update the wavefunction at each time step. This is done by summing the Fourier coefficients times the eigenstates.

```
while True:
    rate(20.0/dt)

    if RSdict['runStop']:
        t += dt
        psi = np.zeros((NA,NA), complex) # initialize psi

        #
        # Here's where you put in your code to compute the
        # wavefunction psi, at later times
        #
        for nmPair in eigenstates.keys():
            psi += coefs[nmPair]*eigenstates[nmPair]*np.exp(1j*
                omegas[nmPair]*t)

        for i in range(NA):
            for j in range(NA):
                SetCylinderFromCN(cylinderScale*psi[i,j],alist[i][j])
```

4.2   SPHERICAL HARMONICS

When considering a 3D system with spherical symmetry the wavefunction can be factored into radial and angular parts $R_{nl}(r)$ and $Y_{lm}(\theta, \phi)$. The radial factor $R_{nl}(r)$ is closely related to the 1D case with an added term that depends on the angular momentum. More on that in a bit, but first let's consider the angular factor $Y_{lm}(\theta, \phi)$, which are called the *spherical harmonics*. These are functions of θ and ϕ that are orthonormal over the entire sphere. The spherical harmonics can be defined, up to $l = 3$, in python like so:

```
#
# Spherical Harmonics
#

Y={}
Y[(0,0)] = lambda t,p: sqrt(1.0/(4.0*pi))
Y[(1,0)] = lambda t,p: sqrt(3.0/(4.0*pi))*cos(t)
Y[(1,1)] = lambda t,p: -sqrt(3.0/(8.0*pi))*sin(t)*exp(+1j*p)
Y[(1,-1)] = lambda t,p: sqrt(3.0/(8.0*pi))*sin(t)*exp(-1j*p)
Y[(2,0)] = lambda t,p: sqrt(5.0/(16.0*pi))*(2*cos(t)**2-1.0)
Y[(2,1)] = lambda t,p: -sqrt(15.0/(8.0*pi))*sin(t)*cos(t)*exp(+1j*p)
Y[(2,-1)] = lambda t,p: sqrt(15.0/(8.0*pi))*sin(t)*cos(t)*exp(-1j*p)
Y[(2,2)] = lambda t,p: (15.0/(32.0*pi))*sin(t)**2*exp(+2j*p)
Y[(2,-2)] = lambda t,p: (15.0/(32.0*pi))*sin(t)**2*exp(-2j*p)
Y[(3,0)] = lambda t,p: sqrt(7.0/(16*pi))*(5*cos(t)**3-3*cos(t))
Y[(3,1)] = lambda t,p: -sqrt(21.0/(64*pi))*sin(t)*(5*cos(t)**2-1)*exp(1j*p)
Y[(3,-1)] = lambda t,p: sqrt(21.0/(64*pi))*sin(t)*(5*cos(t)**2-1)*exp(-1j*p)
Y[(3,2)] = lambda t,p: sqrt(105.0/(32*pi))*sin(t)**2*cos(t)*exp(2j*p)
Y[(3,-2)] = lambda t,p: sqrt(105.0/(32*pi))*sin(t)**2*cos(t)*exp(-2j*p)
Y[(3,3)] = lambda t,p:  -sqrt(35.0/(64*pi))*sin(t)**3*exp(3j*p)
Y[(3,-3)] = lambda t,p:  sqrt(35.0/(64*pi))*sin(t)**3*exp(-3j*p)
```

Note that this approach creates a simple data structure of functions that can be used to pull the correct spherical harmonic from the collection based on the values of l and m as a python tuple.

4.3 ORBITAL ANGULAR MOMENTUM

These quantum numbers l and m are related to the electron's orbital angular momentum. In a classical system the angular momentum could be anything, but in a quantum system the angular momentum is quantized to permit only specific values. The quantum version of orbital angular momentum is made of two parts: the magnitude of the orbital angular momentum L, given by the quantum number l, and the z component of the angular momentum, L_z, given by the quantum number m. The physical values of L and L_z are given by the relationships $L = \hbar\sqrt{l(l+1)}$ and $L_z = m\hbar$, respectively. The magnitude of the orbital angular momentum is quantized to the values $l = 0, 1, 2, 3, \ldots$ and the z component of the angular momentum is quantized to the values $m = -l, -l+1, \ldots, l-1, l$.

4.4 SPIN

The spin angular momentum is a quantum number that is associated with the electron's intrinsic angular momentum. The spin angular momentum is quantized to the values $s = 0, 1/2, 1, 3/2, \ldots$ and the z component of the spin angular momentum is quantized to the values $s_z = -s, -s+1, \ldots, s-1, s$. The physical values of the spin angular momentum are given by the relationships $S = \hbar\sqrt{s(s+1)}$ and $S_z = s_z\hbar$, respectively. The spin angular momentum is a half-integer quantum number, so the spin angular momentum can only take on the values $s = 0, 1/2, 1, 3/2, \ldots$.

4.5   THE HYDROGEN ATOM

The radial wavefunction will depend on the potential energy function, which in a system with spherical symmetry can only depend on r. For example, the Coulomb potential, which is the most appropriate potential to use for a model of the hydrogen atom, produces radial functions that can be set up in simple data structure like so:

```
#
# Some Radial wave functions
#
R={}

R[(1,0)] = lambda r: 2*exp(-r)
R[(2,0)] = lambda r: sqrt(0.5)*(1.0-r/2.0)*exp(-r/2.0)
R[(2,1)] = lambda r: sqrt(1.0/24.0)*r*exp(-r/2.0)
R[(3,0)] = lambda r: 2.0/sqrt(27.0)*(1.0-2*r/3.0 + (2.0/27.0)*r**2)
                     *exp(-r/3.0)
R[(3,1)] = lambda r: (8.0/(27.0*sqrt(6.0)))*(1.0-r/6.0)*r*exp(-r/3.0)
R[(3,2)] = lambda r: (4.0/(81.0*sqrt(30.0)))*r**2*exp(-r/3.0)
R[(4,0)] = lambda r: 0.25*(1.0-0.75*r+0.125*r**2)*exp(-r/4.0)
R[(4,1)] = lambda r: (sqrt(5.0/3.0)/16.0)*(1-0.25*r+r**2/80.0)*r
                     *exp(-r/4.0)
R[(4,2)] = lambda r: (sqrt(0.20)/64.0)*(1-r/12.0)*r**2*exp(-r/4.0)
R[(4,3)] = lambda r: 1.0/(sqrt(35.0)*768.0)*r**3*exp(-r/4.0)
```

The radial functions are defined in terms of the radial coordinate r, the principle quantum number n and the magnitude of the orbital angular momentum quantum number l. The radial functions are normalized such that $\int_0^\infty R_{nl}(r)R_{nm}(r)r^2dr = \delta_{nm}$.

The complete wavefunction for a particular hydrogen atom eigenstate can be constructed by multiplying the radial wave function by the appropriate spherical harmonic to produce the complete wavefunction $\psi_{nlm}(\vec{r}) = R_{nl}(r)Y_{lm}(\theta,\phi)$.

In order to make this easy to do it helps to have a helper class that can construct hydrogen wavefunctions on the fly. Don't worry too much if you're not familiar with classes and object-oriented programming. This last section is mostly meant as a demonstration to support visualizing what these wavefunctions look like in 3D over time.

```
wfs={}

class funcComp:

    def __init__(self, n, l, m):
        self.n=n
        self.l=l
        self.m=m

    def __call__(self, r, t, p):
        return  R[(self.n,self.l)](r)*Y[(self.l,self.m)](t,p)

for n in [1,2,3,4]:
    for l in range(n):
        for m in range(-l, l+1):
            wfs[(n,l,m)] = funcComp(n,l,m)
```

The idea here is that instances of the class funcComp are callable objects that can be used to construct a wavefunction for a particular state. The class funcComp is initialized with the quantum numbers n, l, and m and then the call method is used to construct the wavefunction. The call method takes the arguments r, θ, and ϕ and returns the value of the wavefunction at that point. After the class funcComp is defined, we can construct a dictionary of wavefunctions that can be used to pull the correct wavefunction from the collection based on the values of n, l, and m as a python tuple (n,l,m) as needed.

An associated class HState can be used to create a callable wavefunction with variables r, θ, and ϕ.

```
class HState:

    def __init__(self, n, l, m):
        self.n=n
        self.l=l
        self.m=m

    def __call__(self, r, theta, phi):
        return wfs[(self.n, self.l, self.m)](r,theta,phi)
```

Now, how to visualize all this? You could imagine a 3D grid that displays the value of the wavefunction, similar to what we used in 1D and 2D. Unfortunately, this doesn't extend well to 3D. The number of needed grid points goes up like N^3 where N is the number of grid points in 1D. This quickly becomes impractical.

An alternative is to use the "monte-carlo walker" strategy below that follows walkers around in 3D space and creates or destroys walkers based on the relative probability density associated with the wavefunction at that point. The "size" of the walkers is proportional to the relative probability density at the location of the walker. The details of the algorithm are beyond the scope of this text, but basically each walker is represented by a sphere in 3D. The sphere is colored according to the complex phase of the wavefunction at each location at a particular moment in time. On each time-step all the walkers are given a random "bump" in space. If the probability associated with their new location increases, the update is always kept. If the probability goes down, the new position is kept with a probability equal to the ratio of the new probability divided by the probability of their last location. This is a famous algorithm that's known to produce a distribution of positions, in the long run, proportional to the underlying probability density of the wavefunction.

```
Walkers=random.normal(size=(3,n0))*3+(minX+maxX)/2.0

if AXES:
    c1 = cylinder(pos=vec(minX,0,0), axis=vec((maxX-minX),0,0),
        radius=0.005*(maxX-minX), color=color.red)
    c2 = cylinder(pos=vec(0,1.3*minX,0), axis=vec(0,(1.3*(maxX-minX)),0),
        radius=0.005*(maxX-minX), color=color.green)
    c3 = cylinder(pos=vec(0,0,1.3*minX), axis=vec(0,0,(1.3*(maxX-minX))),
        radius=0.005*(maxX-minX), color=color.blue)

spheres = []

for w in Walkers.T:
    spheres.append(sphere(pos=vec(*w), radius = 0.05, color=color.red,
        opacity=opacity))
```

```
def compPsi(Walkers):
    x = Walkers[0]
    y = Walkers[1]
    z = Walkers[2]

    r = sqrt(x*x + y*y + z*z)      # vector of r values
    phi = arctan2(y,x)             # vector of phi values
    theta = arccos(z/r)            # vector of theta values

    psi = zeros(n0, complex) # initialize psi with zeros

    for n,l,m in coefs.keys():
        psi = psi + coefs[(n,l,m)]*HState(n,l,m)(r,theta,phi)*exp(-1j*t
            *(1.5 - 1.0/n**2))

    return psi

def updateSpheres(Walkers, psi):

    psi = psi/(abs(psi)**2).sum()

    for i in range(len(psi)):
        spheres[i].pos = vec(*Walkers.T[i])
        SetSphereFromCN(scaleFactor*psi[i],spheres[i],maxSize=0.1*diffX,
            minSize=0.001*diffX)


def relaxWalkers(Walkers, n):

    oldPsiSq = abs(compPsi(Walkers))**2      # get old probs

    for i in range(n):
        newWalkers = Walkers + random.normal(size=(3,n0))*ds
                                    # bump walkers a bit...
        newPsiSq = abs(compPsi(newWalkers))**2
        probRatio = newPsiSq/oldPsiSq
        maybeIndexes = flatnonzero(probRatio<1.0)
        testRatios = probRatio.take(maybeIndexes)
        r = random.random(len(testRatios))
        keep = r<testRatios
        oldIndexesIndexes = flatnonzero(1^keep)
        failRatio = len(oldIndexesIndexes)*1.0/len(testRatios)
        if len(oldIndexesIndexes):
            useOldIndexes = maybeIndexes[oldIndexesIndexes]
            newWalkers[:,useOldIndexes] = Walkers[:,useOldIndexes]
                                    # update with rejected Walkers

    Walkers = newWalkers
    oldPsiSq = abs(compPsi(Walkers))**2

    return Walkers
```

Figure 4.2   Hydrogen representation: $n = 4, l = 3, m = 1$, state.

This results in a visualization shown in Fig. 4.2. The grayscale value is determined by the wavefunction phase. The density and size of the spheres is determined by the wavefunction magnitude at the location in space.

4.5.1   Exercise

How does the distribution of walkers change as you adjust the quantum numbers of the energy eigenstate? Pay close attention to the phase as a function of angle around the z axis. How should changing l values change the behavior of this phase? Explain.

Appendix A

INSTALLING VPYTHON

You can install VPython on your computer if you'd prefer that to using WebVPython.

Note: The instructions provided here are current as of the publication of this book, but for the most up-to-date instructions, please visit the VPython website at http://vpython.org.

The current recommendation from the VPython website is to install the Anaconda Python distribution.

The vpython module currently works with Python versions 3.8, 3.9, and 3.10.

The vpython module is available using

```
conda install -c vpython vpython
```

or

```
conda install -c conda-forge vpython
```

or

```
pip install vpython
```

To update to later versions of vpython use

```
conda update -c vpython vpython
```

or

```
pip install -U vpython
```

When running from a terminal, if the program does not end with a loop containing a rate() statement, you need to add "while True: rate(30)" to the end of the program. This is not necessary when launching from environments such as Jupyter notebook, IDLE, or Spyder.

Appendix B

PYTHON CONCEPTS

If you're not very familiar with Python or Numpy this appendix could help you along with an introduction to the basic concepts.

Variables, assignments, and comments

A python variable is just a label for an object:

```
a = 3
```

assigns the object "3" to the label "a". In the future, references to "a" will result in the object "3" being used. If you reassign "a" to a different value, it will forget the old value, and begin to refer to the new value.

You'll also see comments sprinkled in the code to help the learner to understand what's going on. In python a comment is any text preceded by a "pound sign": "#".

So for example we could have:

```
a = 3 # three is a nice number, don't you think?
```

This is functionally equivalent to the statement "a=3", but with commentary that python ignores.

Lists and Dictionaries

A python list is a collection of objects. We use these a lot to organize a collection of things that are related to one another. For example, a collection of arrow objects to display a wavefunction. It's easy to create an empty list:

```
myList = []
```

Once created you can add things to a list by using the "append" method:

```
myList.append(1)
myList.append(2)
```

This will then result in a list with things in it. The command:

```
print(myList)
```

will result in the output:

```
[1, 2]
```

Once can retrieve an item from a list by *indexing* into the list with an integer specifying the desired item (starting with the index '0') like so:

 DOI: 10.1201/9781003437703-B

```
print(myList[0])
print(myList[1])
```

resulting in:

```
1
2
```

A dictionary is also a collection, but rather than indexing using the position of the item, you use a "key" that was originally used to store the item in the dictionary.

Let's create an empty dictionary:

```
myDict = {}
```

Note the only difference between an empty dictionary and an empty list is that we use "{" and "}" rather than "[' and ']". Now let's *store* two values in the dictionary:

```
myDict['a'] = 7
myDict['b'] = 12
print(myDict)
```

This results in the display:

```
{'a': 7, 'b': 12}
```

Note that both the keys (a and b) and the values (7 and 12) are printed. To retrieve a value from a dictionary, just "index" it with the corresponding key:

```
print(myDict['a'])
```

Produces the output

```
7
```

CONDITIONALS

A conditional is a section of code that only executes when a specific condition is true, or false. For example suppose we want to increase x by 1 when t is less than five (t<5) but decrease x by 1 when t is greater than or equal to five (t>=5). We could do this in python using a conditional like so:

```
if t<5:
x = x + 1
else:
x = x - 1
```

Note the placement of the colon ":", and the indention. Python uses indention to specify the scope of a series of expressions. All the code that's indented has the same scope, and executes in sequence. Any code that's not indented is "outside the scope" of the "if" or the "else" clause. You can chain conditions using boolean operators like **and** and **or**. For example, one could also write:

```
if t<5 or t>10:
x = x + 1
y = y - 2
else:
x = x - 1
y = y + 2
```

This would increment x by 1 and decrement y by 2 anytime t was either less than five or more than ten. Otherwise it would decrement x by 1 and increment y by 2.

LOOPS

There are two types of *loops* frequently used in python. The `while` loop and the `for` loop. The `while` loop depends on a condition and continues so long as the condition remains true. The `for` loop iterates over the elements of a list-like object (in our case, that generally means arrays and lists). Here are examples of each:

```
t=0
while t<5:
t = t + 1
print("t=",t)
print("done.")
```

which produces:

```
t= 1
t= 2
t= 3
t= 4
t= 5
done.
```

Similarly one could write a `for` loop:

```
for t in [1,2,3,4,5]:
print("t=",t)
print("done.")

t= 1
t= 2
t= 3
t= 4
t= 5
done.
```

Numpy

Numpy is a library intended to improve support for numerical calculation. We'll be using it primarily because it simplifies mathematical operations on arrays of numbers. One way to create a numpy array is to use the "array" method that takes a python list and converts it into a numpy array. Here's an example:

```
import numpy as np
x = np.array([1,2,3])
print(x)

x = np.array([1,2,3])
print(x)
```

This results in the output:

```
[1 2 3]
```

It looks just like a list! However, it behaves differently. Compare these two behaviors:

```
xList = [1,2,3]
yList = [4,5,6]
xArray = np.array(xList)
yArray = np.array(yList)
print("xList*2=", xList*2)
print("xArray*2=", xArray*2)
print("xList + yList=",xList + yList)
print("xArray + yArray=", xArray + yArray)
```

This results in the output:

```
xList*2= [1, 2, 3, 1, 2, 3]
xArray*2= [2 4 6]
xList + yList= [1, 2, 3, 4, 5, 6]
xArray + yArray= [5 7 9]
```

Note that multiplying a list by 2 gives you a longer list. Multiplying a numpy array by 2 produces an array where each element of the array is multiplied by 2. This is much more useful behavior for us! When you add two lists together you get a list which is the first list and then the second list combined. When you add two numpy arrays together the result is an array where each element is the sum of the corresponding elements of the two arrays. This also works with other mathematical operations, subtraction, multiplication, division, exponentiation, and so on. You can also use the functions in the numpy library that accept numpy arrays are arguments. For example:

```
print(np.exp(xArray))
print(np.sin(xArray))
print(np.sqrt(xArray))
```

produces:

```
[ 2.71828183  7.3890561  20.08553692]
[0.84147098 0.90929743 0.14112001]
[1.         1.41421356 1.73205081]
```

Also, there are some convenience functions in the numpy library that help to create number arrays, like np.linspace, ones, and zeros:

```
np.linspace(0,10,11)
np.ones(10)
np.zeros(10)
```

That produce:

```
array([ 0.,   1.,   2.,   3.,   4.,   5.,   6.,   7.,   8.,   9., 10.])
array([1., 1., 1., 1., 1., 1., 1., 1., 1., 1.])
array([0., 0., 0., 0., 0., 0., 0., 0., 0., 0.])
```

As you can see, `np.linspace` produces an array that spans the range from one value to another with a specified number of elements, while `np.ones` and `np.zeros` produce numpy arrays with either ones or zeros, respectively, with a specified number of elements. Note that one can also use a numpy array as a "list like" object in a for loop like so:

```
for t in np,array([1,2,3,4,5]):
print("t=",t)
print("done.")
```

Python Functions

We spend much time in quantum mechanics discussing *wavefunctions* which are, of course, *mathematical* functions. It's useful to use functions, defined in python, to mimic the mathematical functions that arise in quantum theory. To define a python function you use the `def` keyword. For example we can create a function that takes two arguments, and returns the first plus twice the second, like so:

```
def myFunction(arg1, arg2):
result = arg1 + 2*arg2
return result
```

Then when we call the function with two arguments, we get the desired result:

```
print(myFunction(1,2))
```

produces:

5

Index

Printed in the United States
by Baker & Taylor Publisher Services